*For Matthew*

# Out of the Field
## International Harvester Company Buildings

Sarah Tomac

© May 1, 2020

ISBN 978-1-952265-00-6

© 2020 Sarah Tomac. All rights Reserved.

No part of this book may be reproduced or transmitted in any form of any means, without written consent from the Author and publisher.

Front Cover: International Harvester Dealer Base of Operations building, Millville, Pennsylvania and Original blueprints from Geelong Product Engineering Center, Australia. Both images belong to the Tomac collection.

All images in this publication with the WHS image identification number in caption are used with the written permission of the Wisconsin Historical Society, wisconsinhistory.org in conjunction with Navistar, inc.

*"Prospy," a little farm elf, carries the standard uniform trademark throughout the land by which all International Harvester Company farm machines and implements may be identified showing each and everyone the wisdom of modern tillage and harvesting, spreading Prosperity wherever the farmer employs his words of wisdom and thrift.*

*International Harvester offered everything a farmer needed. Stationary engines, milkers, tractors, trucks, implements of every kind (even some made to order!) "Prospy" was the elf to bring the message to the farmer that it was wise to ease the strain of everyday life for a more prosperous farm.*

*The contents of this book are for historical use and do not reflect current scientific knowledge, policies or practices.*

"IF A THING NEEDS DOING, DO IT, AND DO IT THE **BEST** YOU CAN UNDER THE CIRCUMSTANCES."

**P. G. HOLDEN**

# TABLE OF CONTENTS

| **CHAPTER** | | **PAGE** |
|---|---|---|
| One: | Service Bureau | 11 |
| Two: | Whole Farm | 21 |
| Three: | Houses | 27 |
| Four: | General Farm Barn | 31 |
| Five: | Machine or Tool Shed | 35 |
| Six: | Creamery or Dairy | 39 |
| Seven: | Chickens | 41 |
| Eight: | Corn Crib and Granary | 45 |
| Nine: | Powerhouse | 51 |
| Ten: | Water Supply | 53 |
| Eleven: | Icehouse | 55 |
| Twelve: | Septic | 59 |
| Thirteen: | Dealership | 63 |
| Fourteen: | Factories or Works | 81 |
| Fifteen: | Testing Grounds and Proving Tract | 85 |
| Sixteen: | Engineering Centers | 91 |
| Seventeen: | Twine Mills | 101 |
| Eighteen: | McCormick Camp | 105 |
| Nineteen: | General Views | 133 |
| Twenty: | Means to Purchase | 144 |

WHS #7547 International Trucks lined up outside McCormick Works. Sign reads "International Motor Trucks from the factory Akron, Ohio to Philadelphia, International Harvester Company of America, Highway deliveries relieve railroad congestion."

# INTERNATIONAL HARVESTER
## OUT OF THE FIELD: BUILDINGS

WHS #8360 Cyrus McCormick

The International Harvester Company was everywhere. It is one of the most widely recognized farm and agricultural related companies world-wide - because of one man. Cyrus.

It started with the reaper, and grew from there. It was an invention in a time where industrial revolution was at its beginning. That invention led to improvements, never sitting still in the world of an ever changing demand for better machinery, less labor and higher quality harvest of grains.

School books merely mention his name. History books devote a paragraph or less. Yet this is a man who changed the face of agriculture with one invention. Farms for many years had enjoyed the advancements of metalworking in the implements used to plow, work, till and plant the soil. The harvesting of the grain was left to be done the same way it had for years, by hand.

This passion for constant improvement was ingrained in Cyrus' son, Cyrus H. McCormick. The young Cyrus shared his father's sense of needing to provide the agricultural world with the same prosperity and luxuries that the industry and manufacturing factories enjoyed.

This approach did not end when Cyrus H. McCormick partnered with Deering, acquiring Plano, Milwaukee, and Champion to form the International Harvester Company. In fact, it was strengthened. Each company brought their own pioneering spirit, drive and tenacity together to form one of the strongest and far reaching companies ever in history.

More than a hundred years ago, a majority of the country was full of potential, land was available for a new beginning, unowned by the people and a willingness from the government to encourage settlers; acts were passed - like the homestead act - encouraged many to start west and settle where they may. This opened up the vast prairies for grain production, dairies in the Northwest, corn in the mid-states,

Midnight at McCormick Works.

Total land covered, 229 acres.
Total land occupied by buildings, storage, etc., 153 acres.
Total floor area of buildings, 3,556,603 square feet.
Total length of buildings on basis of one story 25 feet wide, 149,939 feet, or 28 4/10 miles.

Total help employed (average) harvesting machine works, 6,287.
Total help employed (average) twine mill, 841.
Capacity, yearly production, 361,816 machines.
Capacity, yearly production, 29,000 tons binder twine.

Above: Midnight view of McCormick works.
Below: The Deering works.
From the International Almanac 1909; Tomac Collection.

The Great Deering Works, Chicago, Ill.

Land comprising this plant, 76¼ acres.
Total land occupied by buildings, storage, etc., 76¼ acres.
Total floor area of buildings, 2,405,978 square feet.
Total length of buildings on basis of one story and 25 feet wide, 99,024 feet, or 18 7/10 miles.

Total help employed, harvesting machine works (average), 4,098.
Total help employed, twine mill (average), 783.
Total help employed, rolling mill (average), 200.
Capacity, yearly production, 291,454 machines.
Capacity, yearly production, 23,000 tons binder twine.

orchards and groves in the coastal regions and every place was a new challenge, a blank canvas for new beginnings and fresh starts.

As a result, building a farm from nothing was a challenge to some, building as they needed without attention to winds, water flow or aesthetics. Other challenges came from diseases, and the rugged wildness that the land offered as a challenge - animals, fire, floods, and many things that we now understand and are able to prevent or build accordingly.

This understanding comes from trial and error, mistakes made, successes, observations. Using the observations from others and adapting them to suit your own set of circumstances is where the successful farmer thrives.

In a time of inventions and agricultural advancements, the commercial businesses were concerned on how to better educate their customers. While there were many different companies serving their own branch of interest, International Harvester was the most notable company to peruse this.

The direct descendant of this effort is the county extension agent, the home demonstration agent and the different boys' and girls' club leaders. This educational activity for the betterment of all people, especially in the rural areas, set IH above the rest with the aggressive progress of education inside school and out in the early 1900s.

The need for education in the early days of agricultural revolution went hand in hand with improved methods of farming. The homesteading act of 1862 encouraged people to relocate out of town and settle the west.

Homesteading was a way for many families that had met with hardship to start new. Many families that sought adventure outside of their peaceful town surroundings were emboldened to go west. Another incentive to begin in a new location was related to the economic downturn in 1893-1894. Regardless of their reasons for starting a farm, many of these people that settled were unfamiliar with the day to day operations of such a pursuit or they were new immigrants to the Great Country, bringing their knowledge and expertise with them.

These immigrants were affluent in their homeland, yet were not able to establish a farm of their own, separate from the siblings. To partner with the family was not an option as it would not sustain multiple families on one land area. These immigrants often settled in the same area as their neighbors, evident still today with large groupings of Swedes, Germans and Italians in rural areas to name a few.

Inside view of the International Harvester Service Bureau

Above: Making copies of the slides or posters that were available for lectures WHS#6709
Below: The Library of resource, experiments and almost endless information; where letters were received and replied to for the betterment of the agricultural world. WHS#6870

Those that were natives, or from a multi-generation background in the United States, or were removed from the direct agricultural pursuit, needed agricultural education. They needed to know how to do things. Both groups benefited from the assistance that International Harvester Service Bureau provided.

The farm is a factory, a business not to be taken lightly, where the farmer must stay up to date in all the agricultural advances both with equipment and production. Growing enough food for the family, selling the excess for profit and constantly improving the quality of life by making the work easier or more efficient, meant that the farmer was fast becoming the most well-rounded, educated businessman found.

In 1912, IH formed a Service bureau, with the focus on education, efficiency and prosperity. Cyrus McCormick felt that his company was taking too much from the agricultural community and not giving enough back. He also felt that it was not wise to wait until he, or his company, died before making the world a happier place. The second generation to head the harvesting world would help as he wanted - yearly. In the first ten years that the bureau operated, he spent over four million dollars (equal to over $60 million in 2019) with the sole aim of education. Educating whole communities and even whole states to better agricultural practices, such as Arkansas, suffering under the collapse of the cotton industry, became the goal. A majority of the education as a result focused on the southern states and the Corn Belt.

Mr. Perry Holden was appointed Director of the Agricultural Extension Department of the International Harvester Company by Cyrus H. McCormick, remaining until he retired in 1932.

Holden came with an impressive resume.

Born in 1865, in Dodge County, Minnesota, his parents soon relocated in 1871 to Almira Township, Benzie County, Michigan. His father was a surveyor and as a boy he assisted, becoming in his own right an expert. He also worked on the farm and as he grew older taught school in the winter months, as a way of being able to pay for college. He graduated from Michigan Agricultural College (now Michigan State University) in 1889, becoming an instructor in the school from then until 1893. He was awarded the degree Master of Science in 1895 by the college.

Married to Miss Carrie Burnett, of Bancroft, Shiawassee County, Michigan in 1892, who was also a student at MAC. She was one of the first women to attend the college, despite no program available for women until 1896. The couple returned to Oviatt, Benzie County, where he served as superintendent of schools in 1895 and 1896. From there he moved to the University of Illinois, as head of soil physics, and

*Perry Holden on tour, extolling the virtues of healthy crops. WHS #6883*

The Inland Empire Alfalfa Campaign, September 29-October 29, 1919.
The "Inland Empire" consists of the area in the wheat growing region of Palouse, located at the intersections where the states of Washington, Oregon and Idaho meet.
The campaign was conducted for 41 days, over 4,000 miles of railroad, 140 stops made at railroad points; 960 lectures, attended by 86,000 people. nearly 600 lectures were held on farms at peoples homes, as most of the people attending were farmers.
The purpose of the campaign was to urge the necessity of adopting a diversified system of farming in the great wheat region where a one-crop system is in general practice.
This was carried out by the following organizations: Spokane Chamber of Commerce; the agricultural colleges, Washington State Grange and Farmers Union; the state leaders of U.S.. Dept. of Agriculture; the Educational Departments of Washington. Idaho, and Oregon; the Union Pacific Railway; the Great Northern Railway; the Oregon-Washington Railway & Navigation Company; the Northern Pacific Railway; the Federated Commercial Clubs of the Inland Empire; the Editorial Association of the Inland Empire. This was all assisted by Professor P.G. Holden, director of the Agricultural Extension Department of International Harvester Company, with a staff of twelve lecturers.

organized the Agronomy Department, during those four years at the university he also organized the Corn Growers Association and the Sugar Beet Growers Association; assisted in establishing the Corn Breeders Association.

In 1900 and 1901, he was the agriculturalist for the Illinois Sugar Refining Company and in 1902 helped to organize the Funk Brothers Seed Company for the scientific breeding of corn.

He became Professor of Agronomy and Vice Dean of Iowa State College of Agriculture in 1902, again starting an Agronomy Department. In 1906, he also organized and became director of the Agricultural Extension Department of that college where he remained until January, 1912. The ten years at this college was filled full of creation and expansion of many clubs, organizations, and manuscripts throughout the state. Short courses were created and taught to further educate on the study of corn throughout Iowa. His five books on corn culture, which he illustrated himself, have been published throughout the States as well as one was translated into Spanish, for circulation in Mexico, Spain, Russia and Argentine. He was naturally the most well versed person in agriculture in 1912 when he accepted the title of director of Agriculture Extension Department for International Harvester Company.

*Perry Holden; Director of the International Harvester Service Bureau 1912-1932, WHS #8780*

Under his direction, the extensive education he began in Iowa expanded to the world, helping men, women, boys and girls in every state as well as the Philippines, Canada, Mexico, Argentine, Puerto Rico, Holland, Russia and China to name a few.

He retired to Whitehall, Michigan, where his wife and children had been living since 1917. His wife, Carrie, was much celebrated in the area for her care and production of eggs, marketing 20,000-26,000 dozen eggs a year, and selling them above market price. Many families were prescribed by the doctors to consume eggs, and only the best would do. Mr. Holden encouraged his wife the best he could over the years and helped out on Sundays when he was home from Chicago. Her methods of egg production were widely discussed and visitors from many colleges in the nearby states would visit as well to study her method of producing and marketing eggs. The Harvester Company published a bulletin that was based on her work and did much to elevate the poultry business from a few backyard laying hens to a real commercial industry.

One of three I H C Demonstration Farms in the South. This one is located at Brookhaven, Miss.

*From the International Almanac 1909 ; Tomac Collection*

*Example of the guidance Professor Holden gave to many around the world.*
The Arkansas Agriculture Program, 1915, led by Professor Perry Holden, Agricultural Director of the International Harvester Company, taught the state to feed itself.
The campaign was a success because of the full cooperation of the following businesses working together under the direction of the Agricultural College of the University of Arkansas; the four Agricultural Colleges; the State Department of Agriculture; the press; the railroads; telephone and telegraph companies and the communities of which were visited.
The summary of the events to create a united agricultural state, was in four parts.
First, a survey of each county, for the amount of food brought from other states in a normal year that Arkansas could produce profitably.
Second, map out a campaign on these facts.
Third, charts and speakers for the agreed basis, with each county for the speakers to visit, locations and meeting places to be secured as to complete each county in one day.
Fourth, that each local community pay for the expense of transportation, hotels, so as to bring the community into the understanding that it was their responsibility to secure lasting and better results.
The International Harvester Company assisted with thirty field workers, trained in public speaking, the state furnished another thirty, so that sixty trained speakers would be able to teach. Unlimited space, for at least a year after the lectures, were to be given in the major paper (coverage lasted for three years)
Holden insisted that there was to be no advertising of agricultural implements in this, only that the speaker, if applicable, was introduced as being connected to International Harvester and the charts showed "Prepared by the International Harvester Company"
The Harvester Company spent about $30,000 ($755,000 in 2019) for the campaign, and the message to the people "Let Arkansas Feed Herself" in 1915 was a success, as the state found pride in doing better, providing for themselves, and setting an example for many other states.

Perry Holden, a magnetic person, had a way of delivering the most mundane of subjects in an almost magical way, all who heard him speak hung on to every word. In the first seven years at International Harvester, over 17,000 meetings were held for farmers and 51,000 other meetings which were attended by ten million people. Four million copies of 120 different booklets were printed and sent out to anyone who desired a copy. With a focus on education, Mr. Holden and his team did much to advance the agricultural community worldwide.

One of the aspects that IH was concerned about for the farmer was the layout and use of the ideal farmyard. With good reason!
The company wanted the farmer to be happy, and they knew that the way to happiness was for the whole family to be happy and successful. In the 1900s the views on how to have a happy and successful life are not that much different than they are today. Sure there are some ways of delivering the message from then that are different than the ways that happened now. Men and women had clearly defined roles in 1900s. These specific roles varied in nature with the regional and ethnic changes of the farm that was operated. However, Harvester knew that the roles of a woman on the farm were higher demand than other companies. Women were important. Needed. Valued.

International Harvester was not afraid to share that thought with the population in their advertising. Knowing that the woman was raising the most important crop of all, the future men and women of the world, Harvester would remind the farmer in subtle ways in advertising. Using such lines as "any woman or boy can operate a gasoline engine and the cost is not high" or "much attraction for the farm housewife is a running water system...with the labor saved the housewife and the comfort it provides, farm water supplies water for fire protection, irrigation and watering livestock." The company put the importance on the house and the people's lives before drawing attention to the financial reason for installing an engine or building a building or some other form of improvement. International Harvester wanted the farmer to prosper and be productive in their community. A prosperous farmer would, in turn, purchase more equipment from the company that helped him succeed.

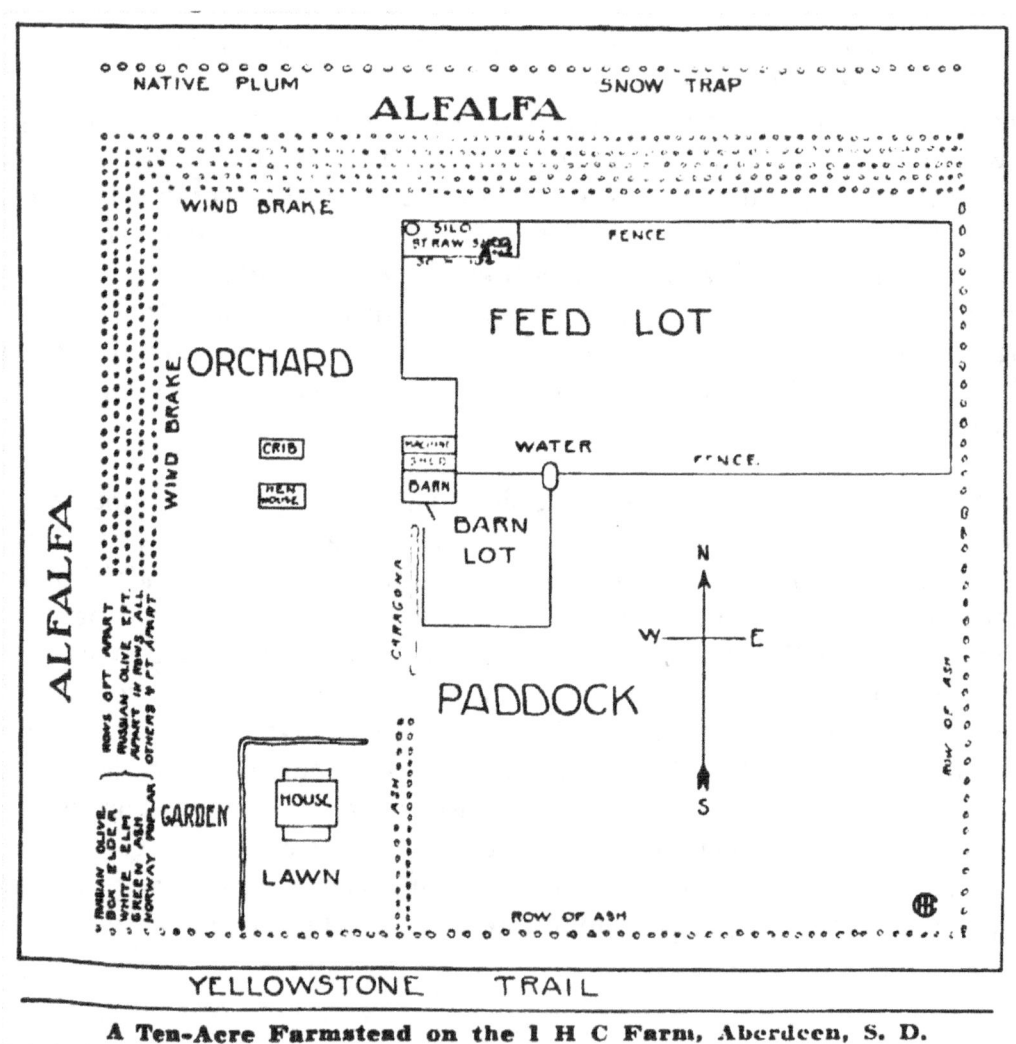

A Ten-Acre Farmstead on the I H C Farm, Aberdeen, S. D.

International Harvester Experimental Farm, Aberdeen, South Dakota.

# CHAPTER TWO:
## WHOLE FARM

Cyrus H McCormick, the second generation to the agricultural entrepreneur, continued what his father had begun. Concerned with more than just providing a quality reaper to the farmers who had lost many sons to the Civil War, he also knew that the people left behind to hold the farm were the wives and daughters of those lost in battle. Creating or marketing the implements to meet the demands that had arisen was simply a matter of good business practice and opportunity.

IH went above and beyond and set themselves apart from other businesses of the time when they also strove to educate everyone - much like a grandfather would to his family - writing letters and encouraging the farmer as if he was one of his own.

Much interest to the profitable and successful farm was based in how it was efficient and organized. The layout of a successful farm was covered in detail by IH and many other farm improvement programs backed by the US Department of Agriculture. There was a country-wide effort to help farmers prosper as a majority of the extra food was sold and shipped overseas. Growing a garden for food, keeping a home, all of these items were covered in detail throughout the years by the service bureau.

An article by J.E. Wing that was published in the 1911 IH Almanac discusses the ideal layout.

He understands that it is difficult to layout the buildings without viewing the lay of the land in consideration, giving suggestions as to the layout, the settings and convenience, stating that "we will not do things well unless we feel that others see what we do."

Ideally, according to 1911 standards, "the house should be set no closer than 100 feet off the road, preferably 400 feet, on a little elevation overlooking the surrounding region and in a little meadow of 2-10 acres. The meadow can be a source of profit as well as beauty when planted in timothy or other alfalfa, away from the house.

The trees should be planted with a definite plan and around the fringes, not scattered about the meadow, leaving a wide open expanse of lawn, much like a lake of grass with trees and shrubs along the shore.

High quality feed is kept in the silo for a healthy and well kept herd of animals, helps production of a dairy farm, like this one above.
*IHC Agricultural Lecture Charts, Livestock on Every Farm, 1915*
Below, a daily view from home, Tomac Collection.

The drive to the house should follow a natural line from the front gate to beyond the house, ideally with a curve, placed with a reason, with trees and shrubs planted as to suggest the reason for the bend in the drive.

There should be two drives - one past the dwelling and to the stable, the other direct from the road to the barnyard, fenced so as to drive animals through it.

The buildings of the barns should be well spaced as to discourage fire from spreading among the buildings. Roofs of slate, tile or metal will lessen the danger of fire. The nature of the farmer determines the final building, for this example we will speak of all the aspects of agriculture. The idea of all the animals, tools, grain, and machines under one roof does not encourage cleanliness and tidiness.

A stable for the horses, another apart from it for the dairy, if kept. One or two cows for milk may be housed in a lean-to at one side or end of the horses. Horses are better in abundant air and cool or cold stabling. Cows for milk do not endure this cold as well. Best results require these to be apart as horses and cows do not get along well as a general rule. An open yard, with concrete if possible, should be provided for both of these animals.

Poultry need separate quarters, as to keep the hay mow and stables free from fowl.

The tool shed should be located in a convenient point as to drive through and unhitch. The simpler the tool shed the better, open at two sides, with posts every ten feet apart and 30 feet deep and as long as you need it for the drill, mower, corn planter or wagon. At one end, place a upper story or bridge that you may let down where you can place tools that will not be needed for months. The beauty of such a shed allows you to drive right though and leave any wagon or machine in place.

 The position of these buildings will need to be determined according to the layout of the land, however there is little economy in having them crowded together. The horse stable deserves a central and easily accessible or convenient location with the carriage house abutting the lawn. It should not be attached to the stable as the smells of the animals and odors of ammonia will fill the carriage and make a most unpleasant ride.

If a dairy is kept, the milk should not be far from the dairy room. The sheep may be positioned farther back and the pig pens should not be where the prevailing winds will carry the odor to the dwellings. For even the cleanest of pig pens have an odor. The poultry house should be positioned in the orchard, convenient to the housewife, as it is her source of interest."

View of the Tomac Homestead 1949

View of the Tomac Homestead circa 1971

A successful farm layout, according to International Harvester, can be found on many farms. Using my own farm as an example, you can see the ideal layout in the early homestead, with the many outbuildings visible. The row of trees line the road to the right of the photo. The buildings were almost set in a direct line, with the large cattle barn, with silo and hay storage at the far northern part of the property. Located next is the milking shed, nestled tight against the cattle barn. The middle of the property is the horse barn, which still stands today as the oldest building on the site. The ability to drive around when needed, with a pasture opening to the south, the dairy house was next.

This dairy is where the milk cans were cooled and stored for the milkman to arrive. Water that was used for cooling came from the well, pumped into the cement holding tank and the overflow fed out to the western side, into a watering trough. Using a series of gates and fencing, the cows always had fresh water this way, no matter which pasture they were located in.

Following farm guidelines, the majority of the grain was stored in near location to the other animals. A granary, corn crib and silo were all located to the east of the piggery, tucked into the trees in this photo. The small building held two or three pigs, raised for slaughter and fresh meat. Finally, the chicken shed is at the southern end of the lineup - near the house, for the ease of egg collection and care.

Farming has changed over the many years on this farm, as viewed in the 1971 photo, the family became focused on row cropping and less animal based. Still visible are the original barns and milking shed, which was used as the repair and storage buildings. Equipment had grown in size, and the only building large enough for the 1456 was the cattle barn, which was replaced soon after this photo was taken.

Diversification is the key to successful farming, being able to adapt as the years pass. Buildings have been rebuilt on the farm in the last few years, adding specific use facilities back again. While there are no chickens, there are still a few market animals raised for fair on the farm. The horse barn is still visible in the final photo, with a couple of wings or leantos added. A large machinery storage shed and workshop replaced the cattle barn and the grain bin replaced the forage silo. The house remains nearly unchanged from when it was built and improved upon in 1880.

Tomac Homestead 2017

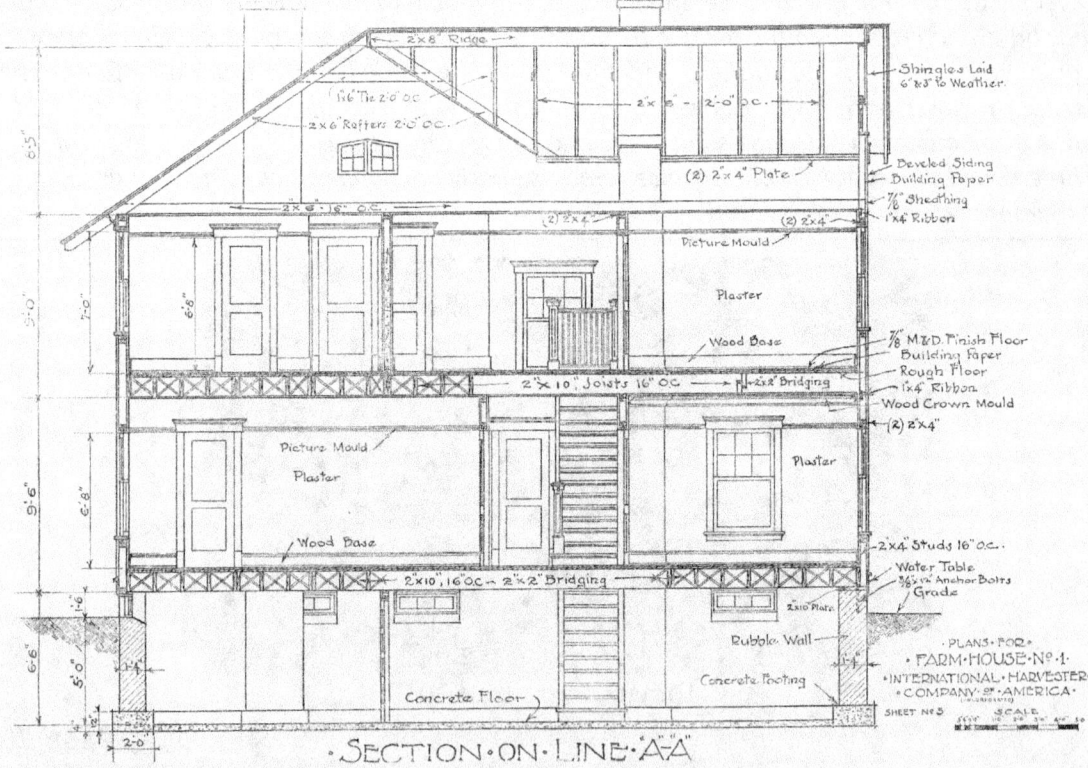

# CHAPTER THREE:
## HOUSES

IH Service Bureau in conjunction with the State Agricultural Colleges, Experiment Stations and the Department of Agriculture was happy to answer any questions that may pertain to farm crops soils, land drainage, farm mechanics or any other problem relating to agriculture that one may have encountered.

The plans for a substantial, attractive, modern, seven room farm dwelling in 1912 are covered in the following as the attention to good barns and other farm buildings were given more attention. The need for a house to be presentable and attractive may have led to securing a future family with less farm bachelors.

The plans from IH were presented and arranged so as to have a desirable presentation for both visitors and least amount of interruptions for the farm help if they were also housed in this main dwelling.

The plans shown in detail are for the larger, two story with basement seven room farm house. The ground floor is arranged so that you may enter from the rear screened porch and go directly up the stairs to the sleeping quarters without passing through any other room of the house. The living and dining room entrance is from the main hall so that to enter or leave it is not necessary to pass through the kitchen or parlor.

On the second floor are four bedrooms that open to a center hall, with access to a bath. From the side bedroom on this floor is a second story sleeping porch, a convenience that many farmers overlooked at the time, as a benefit in the summer months.

The house as viewed should cost about $2,700 to build, based on material and labor in IL in 1912. (Equal to $70,000 in 2019.) The house is finished in clapboards and follows the then modern trend of plain and attractive woodwork. There are no fancy scrolls and no expensive cabinet work indicated with a simple, yet attractive, stairway.

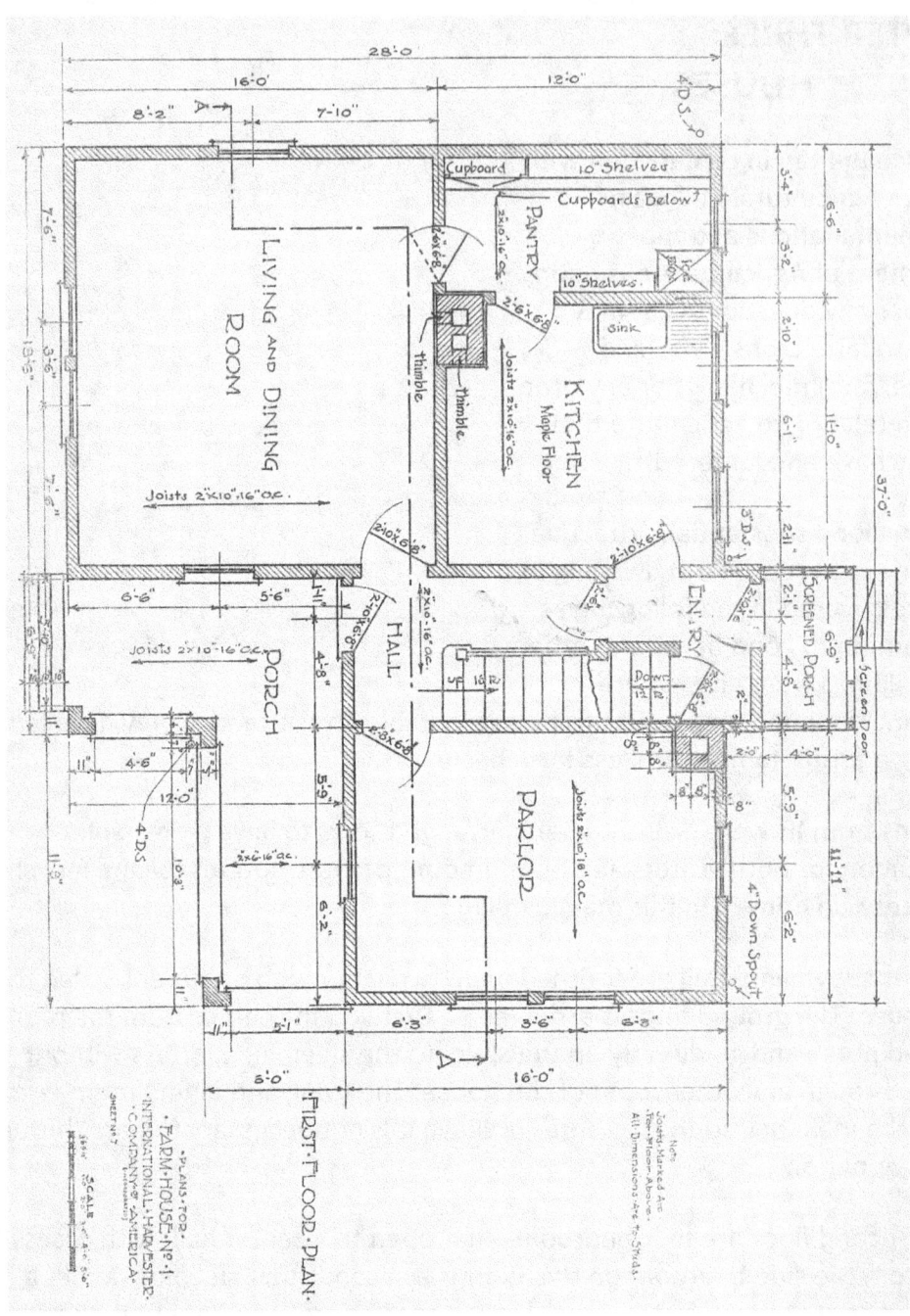

*House Plans shown on these pages are from the Tomac Collection. Some pages not included due to fragile state of plans.*

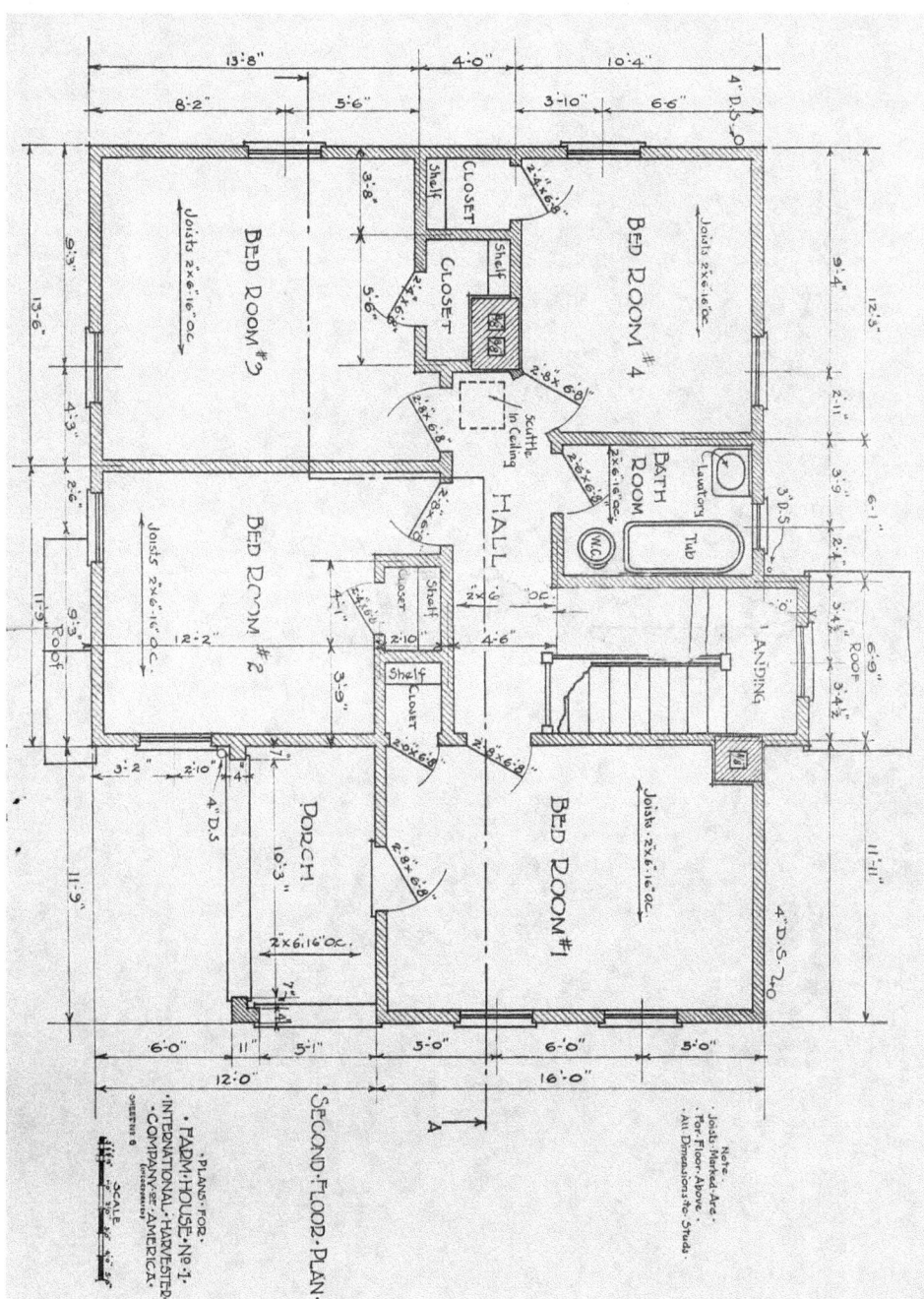

IH issued two house plans. Both of which are an interesting view into farm life in 1912. The plans describe a place for a farmhand to have a room in the house, as was custom of that era to have hired help, or servants. The majority of help to the wife would have been a neighboring girl of teenage years to assist in the care of meal preparations, laundry, child raising, and general house help. The second plan, not shown, is of one more suited to a narrow lot, with the kitchen in the rear, with a more formal setting in keeping with the social standards of the time.

From International Harvester, showing the importance of a Silo on every farm for the well-being of the animals and farmer alike. WHS # 9733

# CHAPTER FOUR:
## BARN and SILO

According to Mr. Perry Holden, in the 1917 booklet about silos, he believes that a silo should be on every farm. It has many advantages for storage of feed through the winter months, offering a very economical way of storing a crop such as corn.

Silos are generally three forms, the pit silo, the bank or bunk silo and finally the most commonly found upright silo. The pit and bunk silo are used more commonly in the areas where the winters are severe, cost of material, labor and transportation is high - areas such as Kansas, Nebraska, Colorado, Montana, the Dakotas and other areas where the winters are severe, the pit silo cannot be blown down or freeze up and they are generally easier and cheaper to construct. The bunk or bank silo is used when there is an abundance of crop, and storage is needed quickly. Previously, corn was shocked in the field; stalk and all, and left out in the weather where it lost up to 30% of its feed value. Using a chopper, it could be harvested and store in the silo for future feed use. Silage was not just made of green corn stalks with immature ears. Silage was made from mature corn with fully developed ears, adding forage from the chopped stalks, and creating a well-balanced ration feed for the production of milk or beef or better sheep. Two acres of chopped corn could be stored easily in silos, ready for winter feeding in addition to alfalfa and other green hay fodder. The larger the farm, the more silo storage is needed. One cow averages about 3 1/2 tons of silage per 180 days, or about 4 cows per acre of feed in 1917. Carefully built silos are an important aspect to the dairyman for well-fed cows will produce the highest quality milk. Round silos have a minimum of surface and wall exposure to prevent danger of spoiling, as round sides reduce the cavities that may form. The most important factor to a good silo is that it is air tight on the sides and well packed to keep out the air which will spoil the fodder. Corn, sorghum, soybeans, kaffir corn and milo maize are the crops most used for making silage, with corn being the best silage crop.

Sorghum contains a large amount of sugar and is recommended to blend it with corn to balance the acidity in the well ripened plant. Clover and alfalfa are good silage crops, but the difference between feeding them dry as hay is small and not commonly used. It is however, preferred when the weather conditions are such that prevents the timely first and last cutting of alfalfa or clover harvest. It is better to place a damaged crop in the silo than to waste it. Both of these plants as well as soybeans are high in protein and can be fed with corn silage to good advantage.

Many different materials can be used to build a silo such as cement, cinder blocks, cement staves, glazed tile, wooden staves, bricks, and so on. The essentials of a well-designed silo for any type of building material used need to follow these guide-

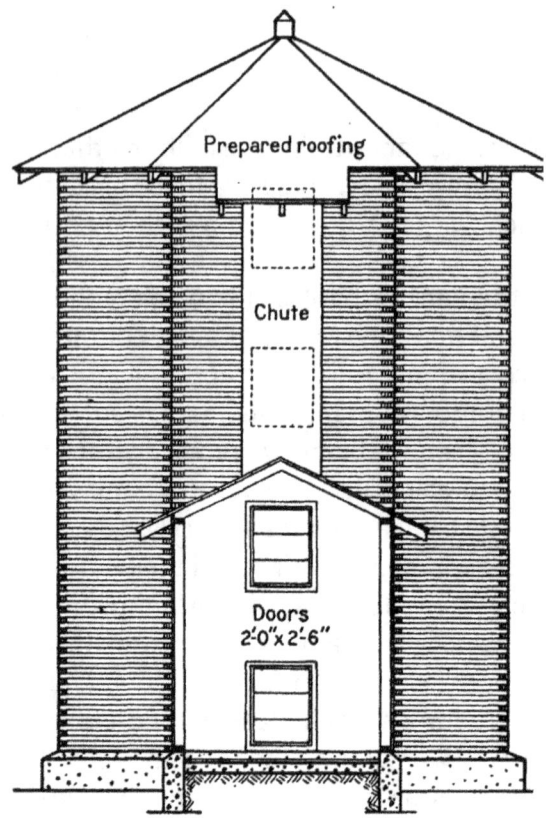

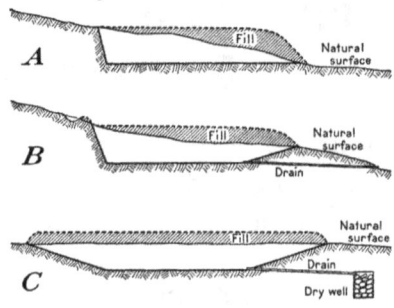

Silo Drawings from various International Almanacs.

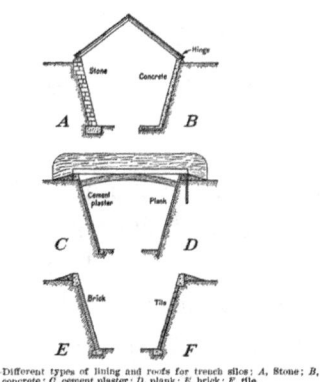

—Different types of lining and roofs for trench silos; A, Stone; B, concrete; C, cement plaster; D, plank; E, brick; F, tile.

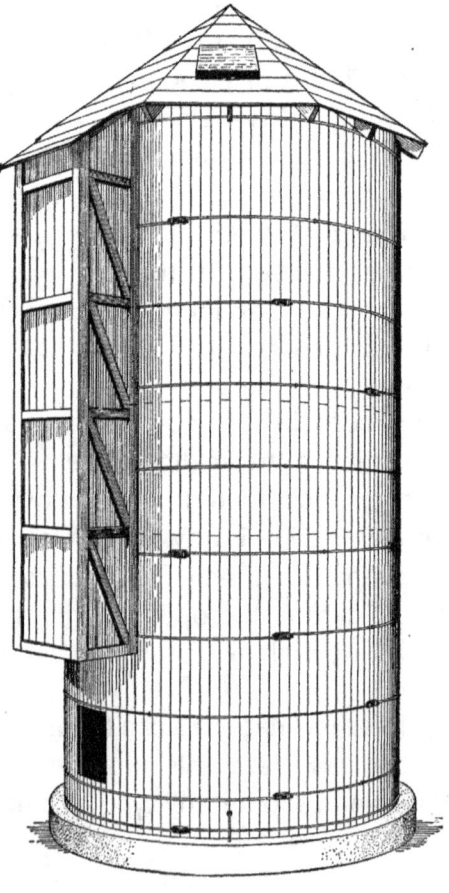

lines.

1. The wall should be practically air tight
2. The inner surface of the wall should be smooth and perpendicular
3. The inner surface of the wall should be free from corners. Round silos are more efficient and economical than other types (square and octagon the most common at the time)
4. The wall should be sufficiently non-conducting to prevent excessive freezing - especially when the silage is to be fed during cold weather.
5. The wall should be sufficiently firm or sufficiently well anchored to prevent cracking due to settling or racking due to wind.
6. The doors should be so designed that a minimum amount of silage has to be removed before they can be opened.
7. A good ladder should be provided with steps from 15"-18" apart and at least 3 1/2" away from the silo or walls of the chute.
8. The foundation should be heavy, well made and reach below the frost line
9. A good roof makes the silo more durable, adds greatly to the appearance and it tight assists materially in keeping the silage form freezing.

A well laid out barn will find the silo conveniently located in near vicinity of the animals to be fed.

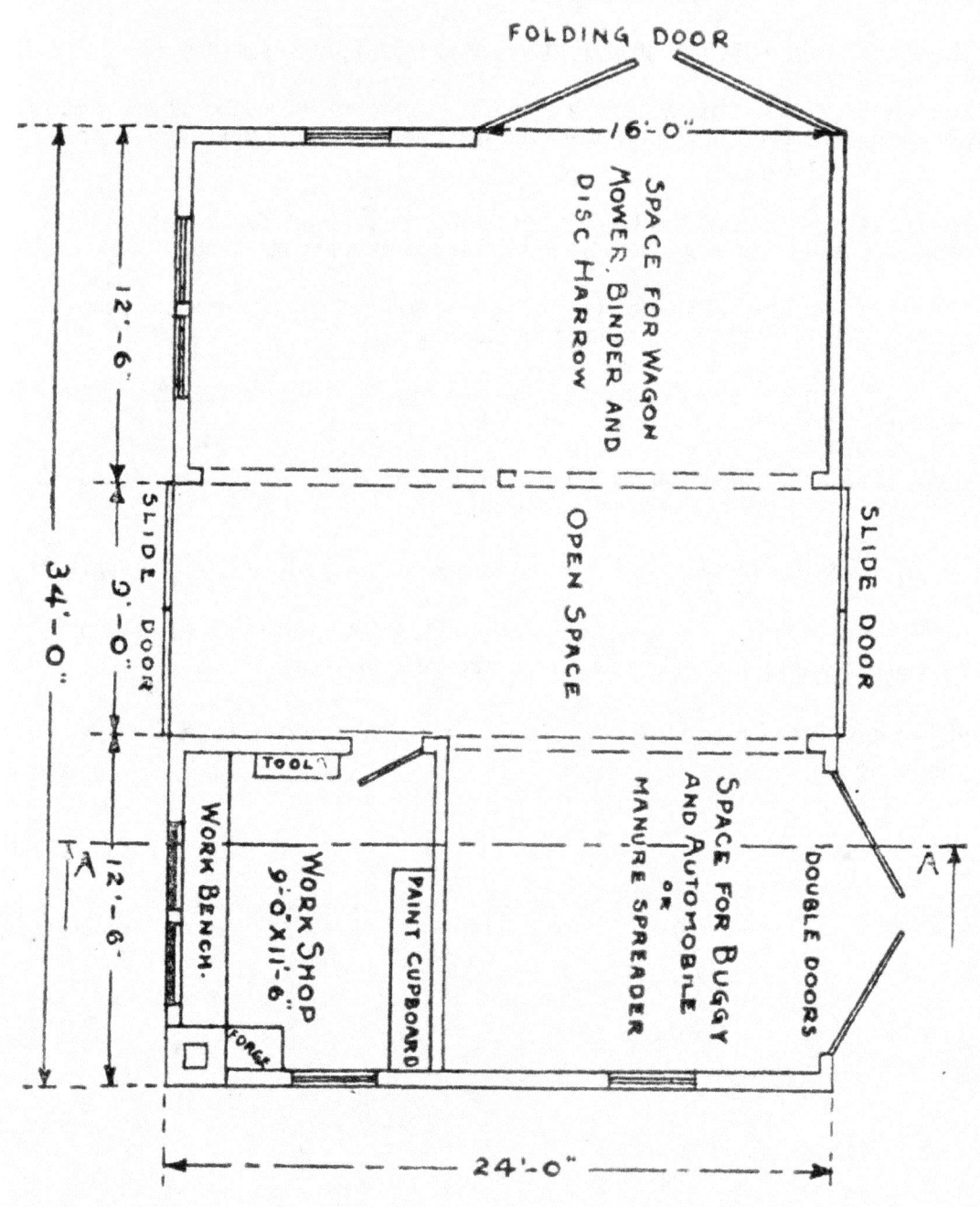

FROM THE INTERNATIONAL ALMANAC; TOMAC COLLECTION

# CHAPTER FIVE:
## MACHINE or TOOL SHED

The first step in caring for your farm machines was to have a suitable space for housing. The space needed depends on how they are arranged. Parts may be removed, reducing the space required, although many machines require a fairly small space to see that they are all housed. This will improve the lifespan of the equipment and reduce the unnecessary repairs during the valuable season.

The IHC Service bureau has provided some sketches for a proposed machine shed that will be both suitable and attractive. Finished with attractive drop siding and shingles set on concrete piers, this is a most suitable shed for any farm. With a loft above for storage or smaller implements, the ground floor remains clear for the ease of movement around large machinery. A convenient workshop set in a corner allows for all tools and paint to be safely and easily used. A small forge located in the corner of the work shop allows for everyday repairs. The buggy and manure spreader or automobile is accessed by double doors to the same side as the workshop. The large open space in the center allows room for those days of inclement weather, where you can unharness the team out of the weather.

To the side is a space for storage of all the implements and machinery, with access from outside as well as the center. This building can easily be adjusted to meet storage requirements by lengthening the area in either direction.

The machine shed drawn is 34 feet by 24 feet. An reasonably adequate size in 1912, it is far too small for our farming needs over one hundred years later. Although our building sizes have grown with the size of the equipment, the principle of housing all our machinery and making repairs to lengthen the lifespan of our equipment remains the same.

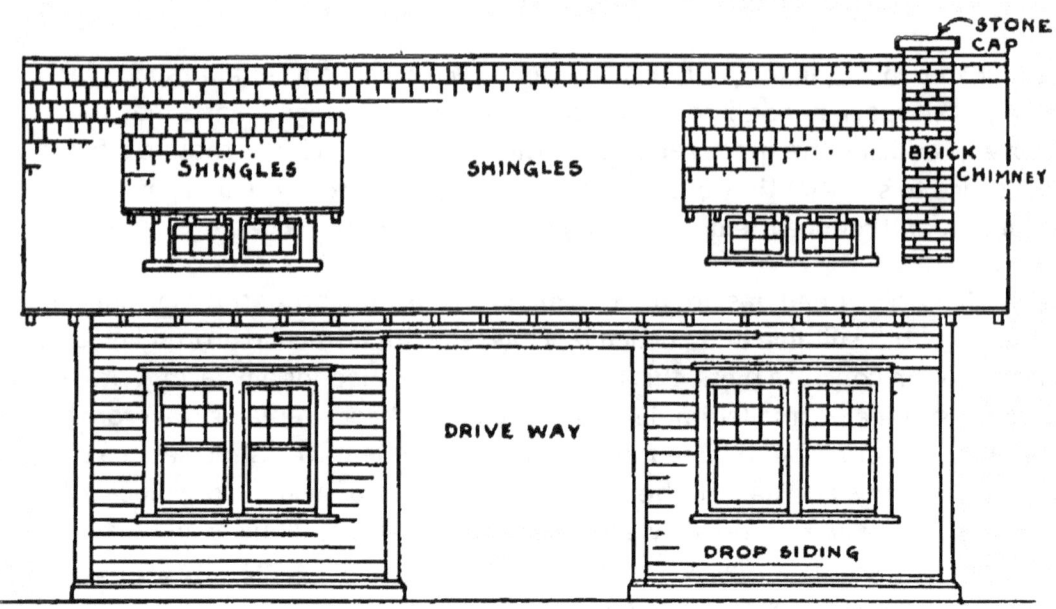

FRONT ELEVATION

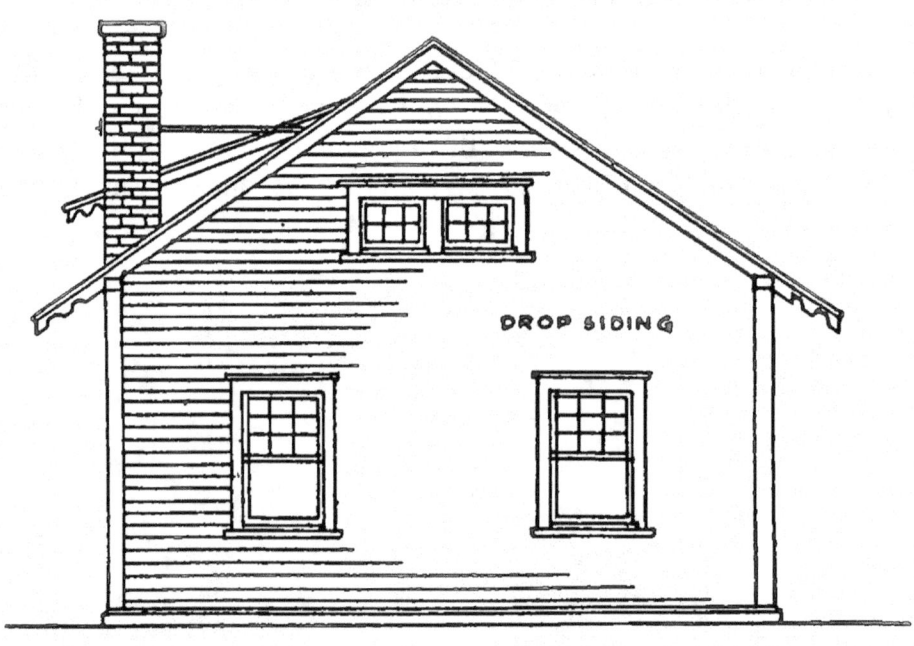

SIDE ELEVATION

From the International Harvester Almanac 1917; Tomac Collection

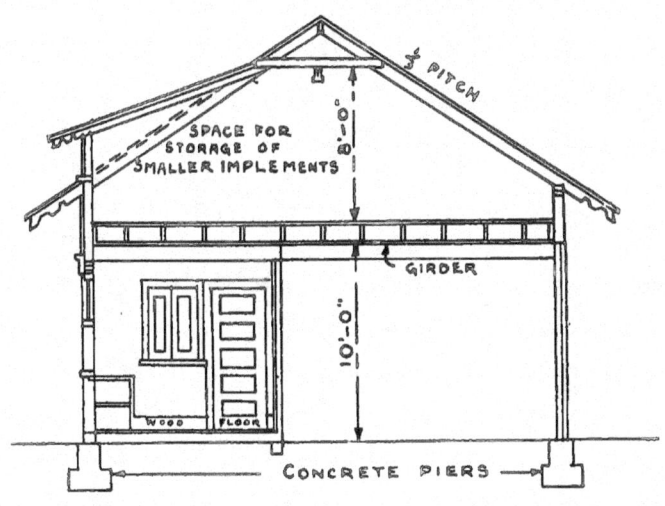

Sketch of Farm Machine Shed
IHC Service Bureau
International Harvester Co of America.

## Space Required to House Farm Implements and Machines

| Implement | Floor Space | Height |
|---|---|---|
| **Plows** - Walking | 8x3 | 3 |
| -Engine Gang | 16x7 | 5 |
| **Harrows** - Single disc, 8ft with truck (tongue off) | 6x9 | 4 |
| **Grain Drill** - 12x7 (add 10ft for tongue) | 5x8 | 5 |
| **Roller** - Corrugated, double | 4x8 | 4 |
| **Corn Planter** - Two row, (add 10ft for tongue) | 5 1/2 x6 | 3 1/2 |
| **Cultivators** - Riding, two row (add 10ft for tongue) | 8x10 | 4 |
| **Mower** - 6ft, (add 10 ft for tongue) | 5x6 | 7 |
| **Binder** - Grain 8ft. (tongue truck off) | 8x14 | 7 |
| - Corn (tongue off, add 11ft for tongue) | 13x6 | 7 |
| **Thresher** -32inch | 26x8 | 8 1/2 |
| **Kerosene Tractor** 8-16HP | 12x6 | 7 |

Hay Rake, Tedders, Hay loader, Ensilage Cutter, Corn Shredder, Corn Sheller, Hay Press, Fanning Mill, Wagons, Buggies, Grain Dump, Potato Planter, Potato Sprayer, Potato Digger, Manure Spreader, were also listed, as well as other variations on implements already mentioned.

Space is listed in feet

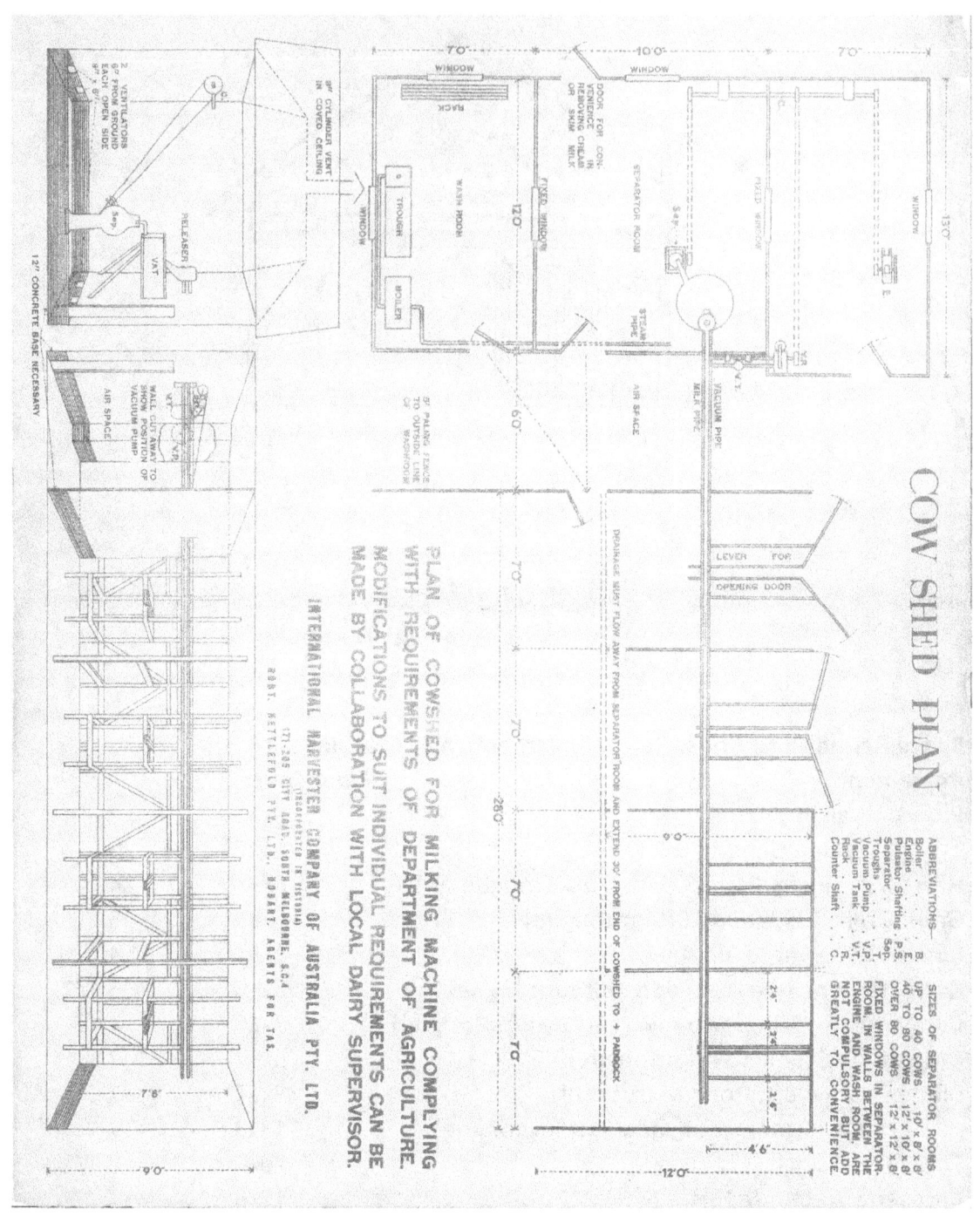

Plan of a Cow Shed in Australia, suitable for a small herd. *Tomac Collection.*

# CHAPTER SIX:
## CREAMERY or DAIRY and COWSHED

When cows are kept and butter or cheese is made, a good dairy house is necessary. The size of the building must fit the size and nature of the business. A steam boiler is essential for heating and cleaning milk cans and utensils. A type of power is also needed for driving the butter churn and operating the pump.

The building should have a cement floor with well constructed drainage.
A benefit to working the milk at home is feeding the skim milk while is it warm from the separator. The pigs and other animals will benefit from this. When cheese is made, whey is not a valuable food product, however it is worth a little in connection with pasture and grain. The plan shown makes a helpful little butter factory for the farm where fifteen to twenty cows are kept.

This Page: A view of an attractive dairy house, and floor plan.

The cowshed shown at left is an overview a small milking cowshed with room for processing milk and operating the machinery in an adjacent building, distributed by the Australian branch of International Harvester.

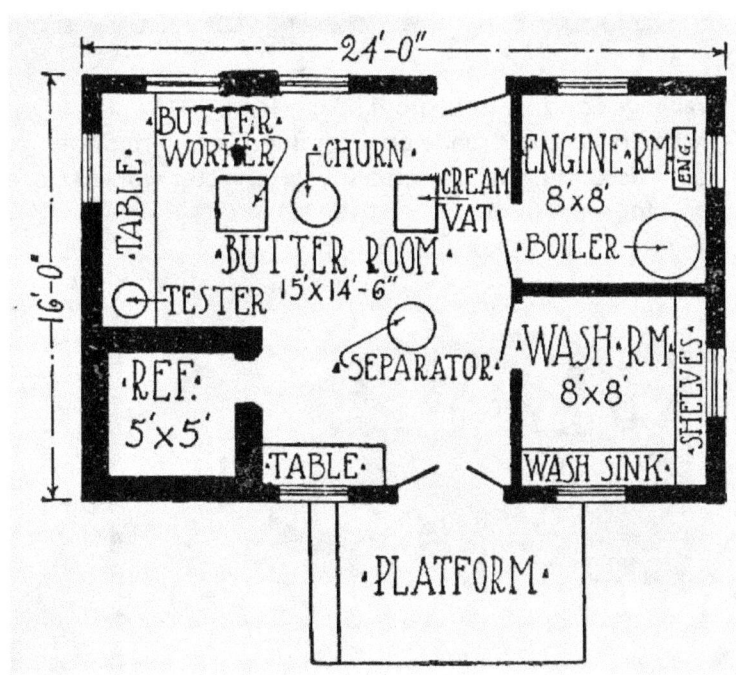

**PAGE 39**

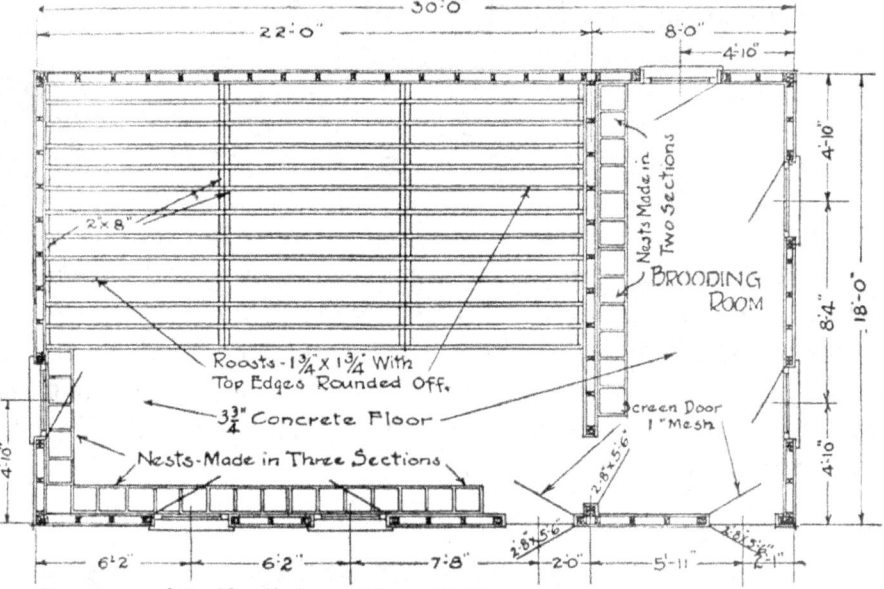

Drawings Courtesy of Radford's Farm Plan, 1915.
These drawings are made on recommendation from the Agricultural Experiment Station, Wisconsin, which international Harvester worked closely with in regards to ideal plans for every farm.

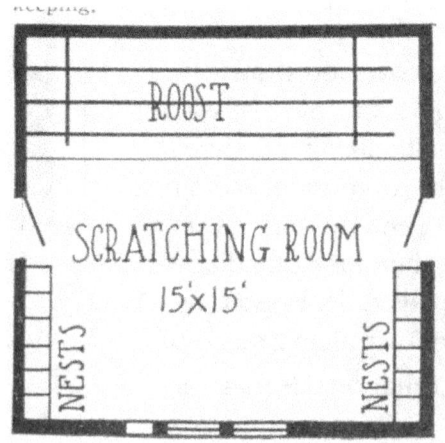

PAGE 40

# CHAPTER SEVEN:
## CHICKENS

Many times, poultry on farms was viewed as a general sideline on the farm, a source of breakfast. The Harvester Service bureau was once again, determined to help the family farm become profitable and productive in the rural areas as they knew that every little bit of income generated on the farm helped to increase the profitability of the whole community.

Chickens are great for keeping bugs and worms at bay and cleaning up table scraps or left over grains on the farm. In 1917, IH compared the income from farm chickens to the greatest feat of the day - the Panama Canal. At at cost of $300 million to complete, the poultry business combined from farms was enough to build two canals, or an annual combined small farm worth of $600 million. 90 percent of the poultry income came from the small family flock. The flock that the wife and children kept for a few eggs and meat on occasion. The same poultry flock that allowed the farm wife to buy material for making clothing, shoes, birthday and Christmas presents, and when the weather was especially unfavorable to the farmers crops, this pocket money would also make the payment for the farm or machinery.

Chickens need good housing, proper feeding, better handling and marketing, as well as an improved flock for the most amount of return.

A great house for the hens did not mean expensive or difficult. Educating those that raised poultry meant teaching the basic needs that the chickens need. A chicken coop needed to be: well ventilated, free from drafts; clean; free from vermin; convenient; comfortable; light and dry. The building should be located facing the south with the north, east and west sides wind proof. Not too close to other buildings, near a grass yard and easy access from the house. Easy to clean, and tall enough for the owner to be easy to work in. Not too narrow as it will be too difficult to ventilate. Well lit, as sunlight is a way to destroy germs and keep the coop dry. It should be draft free and well ventilated, free from bugs, mites and lice, and protected from outside predators.

Concrete foundations and floors are rat proof and easy to keep clean, with a wide door for feed can be wheeled in and manure wheeled out. Height should be no more than 7 feet at the front and 5ft in the rear, tall enough to easily clean and keep warm. For small flocks, four to five square feet per hen and large flocks, three to four square feet per hen.

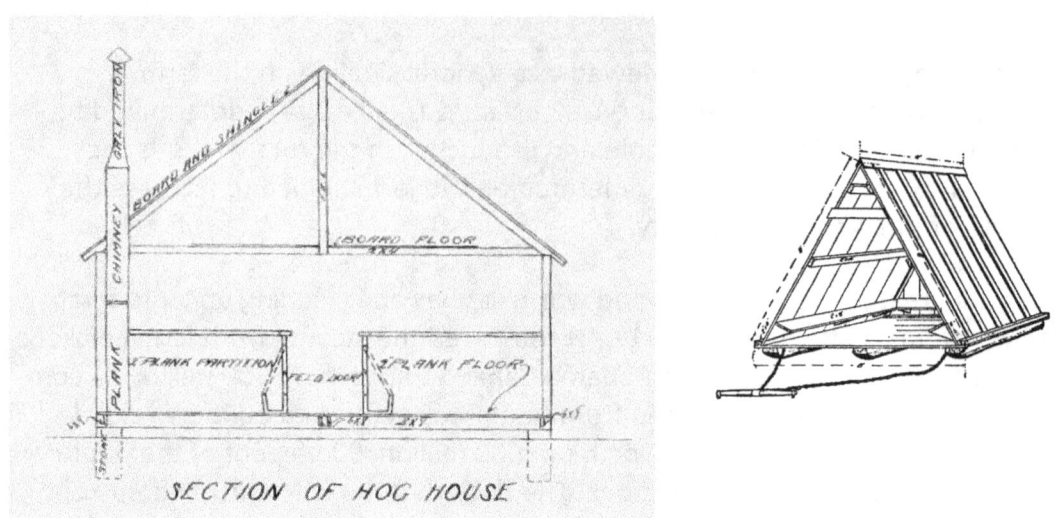

Drawings show the ideal layout for profitable pig raising.

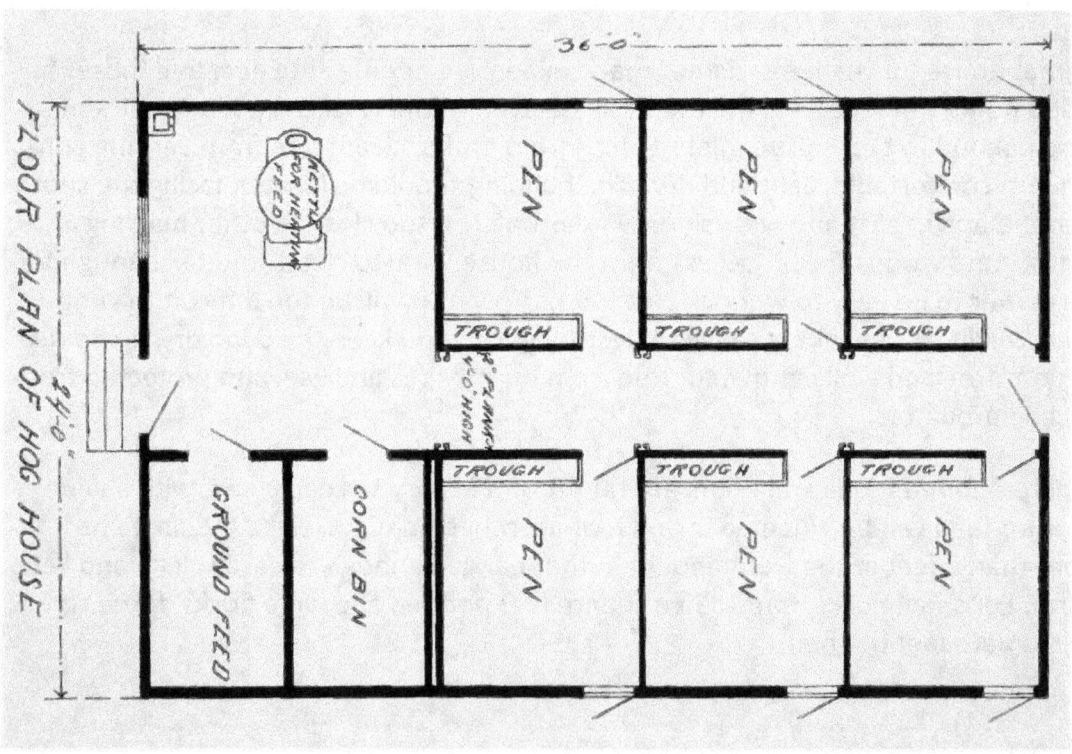

A guideline of dimensions are as follows: 25 hens, 12 feet deep by 10 feet long; 75 hens, 16 feet by 10feet; 100 to 150 hens, 16 feet by 32 feet; Longer houses should have partitions in them to stop them from being drafty.

The front should be open, to allow fresh air and light in. Doors and windows should be arranged so that every part of the floor will be flooded with lights as some point in the day. Long narrow windows are better than square ones. Build perches on the level or hens will fight for the highest and may be hurt. One nest for every four or five hens is sufficient.

Paint the outside of the hen house, make it look as if it belongs to somebody. A cement walk is appreciated by the lady of the house, and helps to encourage pride, cleanliness and prosperity.

## PIGGERY or HOG HOUSE

Keeping a pig is a great source of lard, meat and income throughout the year. One litter of pigs is 7 or 8 fattened hogs that can be sold off, and one sow kept from that litter will grow the herd to two sows with litters, growing the business quickly. Pigs are fairly easy to feed as they will forage for food as well as clean up scraps around the farm and table. The plan at left is from a state agricultural extension recommendation. Harvester had such plans available as well, recommending a clean cement pool for pigs to stay cool and healthy.

Needing shade in summer and a warm place in winter, the IH Service Bureau recommended an "A" shaped building on runners that could easily be moved around for only one or two pigs in the orchard.

A larger Hog house was ideal for the farmer who chose to make a profit and living from raising and selling animals.

Exterior of Chas. Griesemer's Concrete Corncrib at Hopedale, Illinois

From International Harvester Almanac, 1914, *Tomac Collection*.

Interior View of Mr. Griesemer's Concrete Corncrib, Showing Recess in Floor for Drag Belt Conveyor of Corn Sheller.

# CHAPTER EIGHT:
## CORNCRIB AND GRANARY

The threat of fire was always present in the early days before electricity. Kerosene lanterns, candles, open flame of any kind that shed light in the darkness was not to be trifled with. In 1914, International Harvester Company Service Bureau wrote an article on the construction of a concrete corn crib and the added value of such.

The article brought to light Charles Griesemer, an enterprising farmer in Hopedale, Illinois, who built his own concrete Corn Crib. The structure itself was a double crib style, with each crib 10 feet wide and 40 feet long, 14 feet to the eaves and 25 feet to peak of roof over the 10 foot driveway in between. Additional storage bins begin at ten feet over the driveways the length of the building. Each of the cribs holds 1,800 bushels of ear corn. Using a dump elevator and filling to the roof and over the driveway, the capacity if the cribs is increased to 5,000 bushels.

The design of a concrete building over a wooden one prevents worry of fire, reduces the work and expense of painting and repair as well as not having to rebuild one every ten years or so.

The ventilation of the concrete crib has as much as the old wooden style, which was sided with 1x4" boards spaced one inch apart. The block was 24x10x8" with 7 air holes the entire 8" thickness. The hole size of 1 1/7" x 6", spaced two inches apart and two inches from the top, bottom and ends of the block.

The 10 inch foundation walls were built in the usual manner, 3 foot deep and 18" above ground level. The center of the crib floor a cemented trench was laid, 16" high and 22" wide, for the drag belt conveyor of the corn sheller. At crib level, this space is covered with short boards, and removed one at a time during shelling, allowing corn to drop into the conveyor.

Simple, Single story granary with platform for efficient loading and unloading with floor plan, 1912.

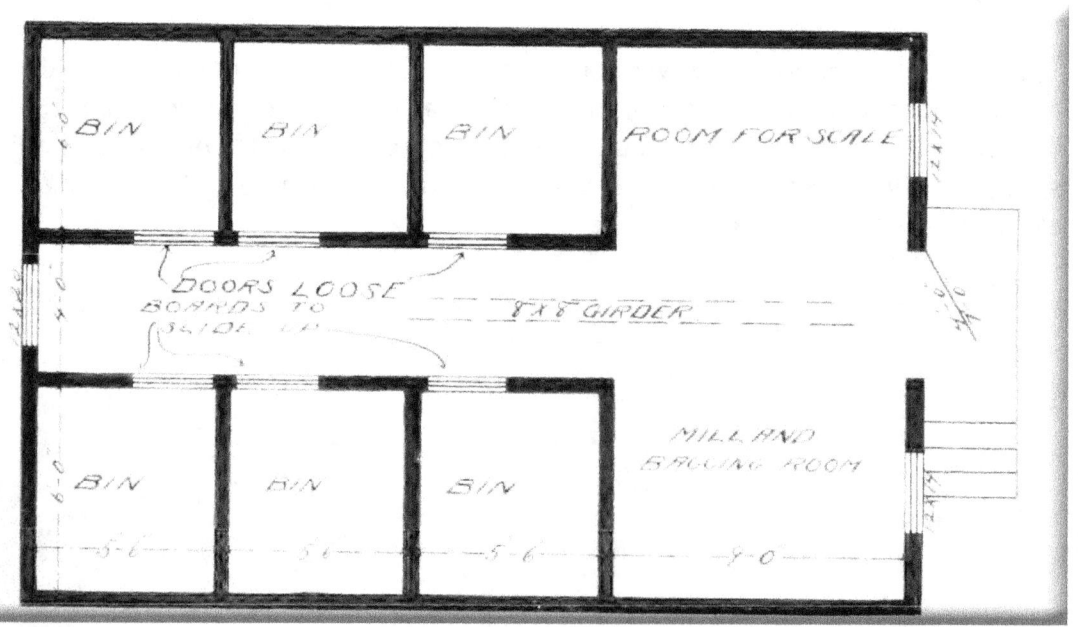

# GRANARY

A farmer in 1912 needed to store grain for the upcoming year of planting and feeding to his animals. He also needed a place to store his grain that he planned on selling during the winter months, when the demand was higher and he could gain a higher profit.

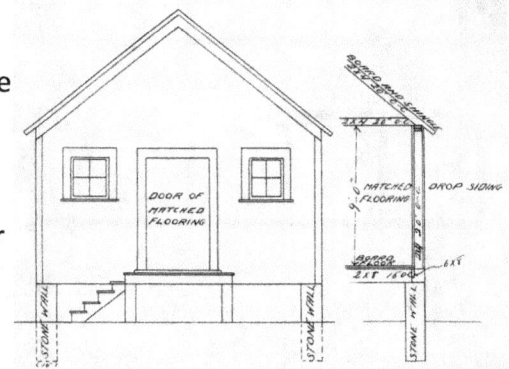

A small building, usually referred to as a granary or seed house, set apart from the main animal barn and near a corn crib with access to a scale house if the farmer was inclined to sell feed to his neighbors served this purpose. The building needs to be rat proof, vermin proof and watertight. Built at least 18" above ground, ideally so that the floor of the granary is the same height as the wagons for easy loading and unloading as well as ventilation under to keep the rats from burrowing in and excessive moisture to rot the stored grain.

The inside is laid out in a way that multiple grains are stored and kept for seed. There is a clear area that allows for a seed grader, keeping the heavy, good seed for planting the next crop and the remainder sacked and kept for feed to the animals. The doors are loose boards dropped into grooves so that they may be put in or removed as required.

The bins would extend to the roof so there is no loft or ceiling in this small purpose built building. The frame of the building is built in the usual way with studding, however ceiling boards (tongue and groove) are nailed to the outside of the studding, and the clapboards are nailed over the tongue and groove boards. The boards must be of good quality for this purpose and carefully put on. A window on either end allows for air movement when filling the building during harvest, and must remain closed to keep out vermin of the air.

A wide door that opens onto a platform is useful for loading wagons and handling bags.

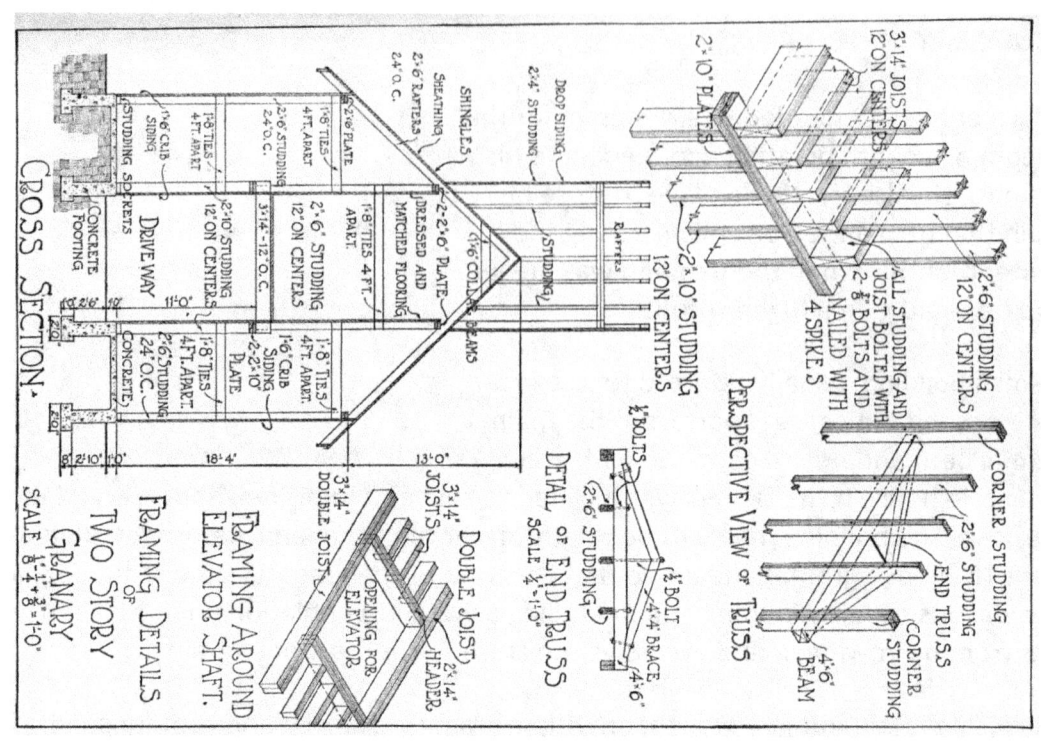

Floor plan and cross section view of a two story granary

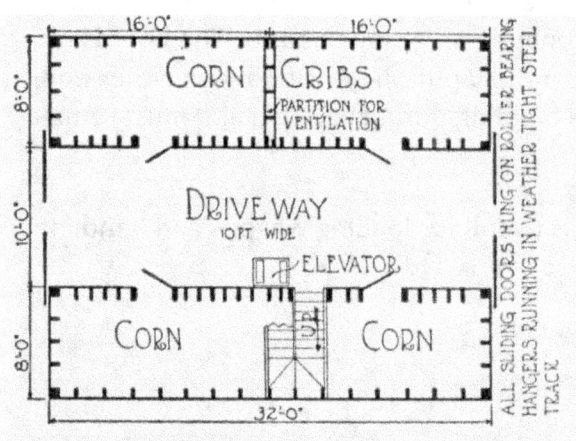

Ground Floor Contains Center Driveway and Two 8-Ft. Corn Cribs.

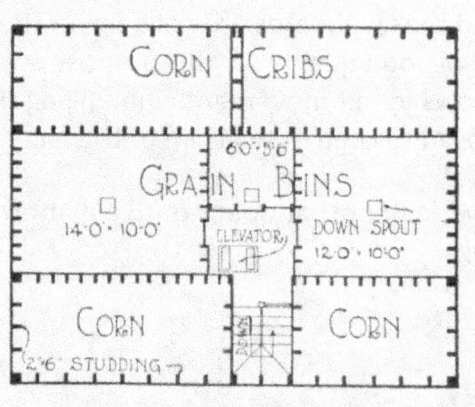

Upper Floor Has Corn Cribs on Outside and Grain Bins in Center.

# MODERN GRANARY AND CORN CRIB COMBINED

A granary and corn crib combination may be better suited for the modern farmer who takes advantage of the machinery available to help with efficiency and prosperity. When building such a modern two story building, care and consideration must be exercised when planning this labor saving building.

There are different equipment needs to consider before undertaking a building, most importantly head-space, type of elevator and power. Ample head-space is needed in the driveway to elevate the wagon when emptying. Stationary elevators operate with buckets in a vertical shaft, filling at the bottom and emptying at the top into a chute directed to the designated grain bin. Space is needed in the cupola to hold the elevator head with room to swing the spout from which the grain discharges. A portable elevator works on an incline and pushed the grain or ears of corn through a trough or spout by means of sliding cross arms. Certain makes of elevators require a pit for the grain to empty into and others have a folding style. Also, one must consider horsepower. There are two kinds of power to drive the elevator - gasoline or oil engine. Great care should be taken when placing the power, generally recommended to house the engine in a separate area near the elevator. In the plans shown, you can place the stairs near the elevator with care to leave room for the engine to operate underneath, with room around the engine for daily oiling and if needed, repairs.

A cubic foot of wheat weighs 49lbs, corn 44lbs, oats 28lbs, peas 50lbs. Figuring the weight of wheat, a bin 10 feet deep filled with wheat will put a load of 490lbs on each square foot of floor surface. If the alleyway through the corn crib is 10 feet wide and the grain bins the same size overhead, with the floor joists placed a foot apart on centers, a full bin 10 feet deep with place a load of 4,900lbs on each joist, approximately two and a half tons. These figures give an idea of what is needed in the construction of a two story grain house to carry the necessary load.

An I H C Engine installed to provide a practical Farm Power House

Power houses from International Harvester Almanac, various years, *Tomac Collection*

# CHAPTER NINE:
## POWERHOUSE

In the almanacs and booklets sent out to its readers, IH mentions several times in various ways how a stationary engine is important to the farm. In one of the more detailed descriptions in 1909, it describes the house as a three room building, with a overhead line shaft.

"The engine room should be kept separate from the rest of the building, well-lit and well ventilated. In this same room should also contain a heater for the winter, a grindstone, emery wheel, drill, and workbench, creating a small workshop for convenience. The gas tank is recommended to be buried underground about eight to ten feet away as a insurance against fire risk, and kept cool in summer." (Over one hundred years later, this burying of the farm gas tank is not recommended.)

The center room housed the grinder, sheller, cutter and fanning mill. This room needed to be well-built to contain the dust that the machines created. The room furthest away from the stationary engine held the churn, cream separator, washing machine, pump and drain, dairy table and a large washing trough, becoming the dairy.
The line shaft could be extended through the outside of the building for a blacksmith bellows, bandsaw, circular saw or any other wood-working machine to be used with the same power.

In 1913, the Almanac reviewed their advice and drew a slightly different example. This shows the engine room and the dairy as two rooms in one building adjacent to the barn where the sheller, grinders and other dust producing machines would be located away from the delicate nature of the need for cleanliness in the dairy.

Many different plans and examples were described over the years, simply because the different agricultural pursuits required different approaches. One thing remained clear however, and that was the need for tidiness, cleanliness, efficiency and economy, leading to great prosperity.

Fig. 1. The Elevated Tank System

The figures illustrate the gas engine driven pumps in both elevated tank and pneumatic tank system.

Fig.1 The elevated tank. The water is pumped from the well into the elevated tank which is higher than the highest discharge point on the farm so that there is ample pressure to all water points throughout the farm. This outside tank would do better in the loft of a barn or attic of the house in the areas where it is subject to freezing.

Fig.2 The pneumatic tank. The tank is usually located in the basement or buried underground where there is no danger of freezing in winter and water will be cool in summer. The tank is air tight and is pressurized, forcing the water up to the points of discharge.

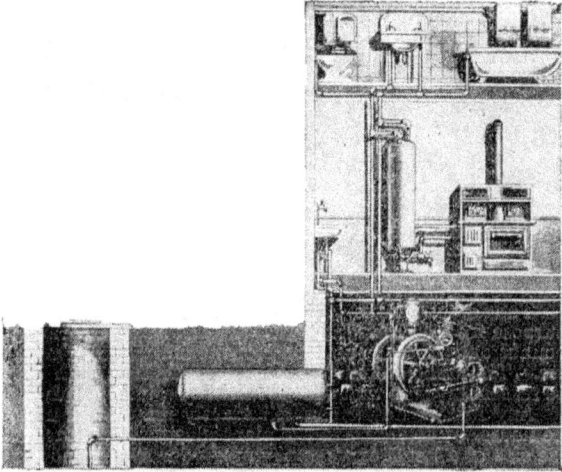

Fig. 2. The Pneumatic Tank System

# CHAPTER TEN:
## WATER SUPPLY

The one feature to reach many rural houses long after their city-dwelling neighbors was the ability for a running hot and cold water system in the house. In addition to the labor saving private water supply, it added much convenience to the daily farm use as well. Not only does the water supply provide water for fire protection, irrigation, or watering the stock, the running water available in the stable saves the farmer labor and time as much as it saves his wife in the house.

On pure business proposition alone, a running water system saves valuable time and adds much to the pleasure of farm life. No modern farm should be without a private water system much like the farmer should not have to harvest wheat with a cradle or cultivate his corn with a hoe.

Two general features need to be decided on when planning a water system;
1. the type of pump to use; whether hand, windmill or power operated.
2. Method of distributing water to the various points needed, whether from an elevated tank which the water flows by gravity or a pneumatic tank installed below ground and forced up by air pressure.

With a small farm and small family, a hand pump is satisfactory, filling up the tank each morning for the daily use. When the family is large, or if water is required for the barn, trough and other service, a windmill or power pump should be used. A disadvantage to the windmill is a large tank must be provided to hold the water as well as it cannot turn when there is no wind, and the water is not supplied fresh everyday. The power pump is the best outfit for all the farm and water can be pumped whenever it is wanted. The cost of operating a stationary engine is not high and any boy or woman can operate it. The engine can also be mounted on a truck to move about to assist where needed when not needed for pumping.

Both of these systems the hot water is obtained by piping that flows through the firebox of the kitchen range, heating the water and depositing the water into a storage tank for use at the hot water faucets.

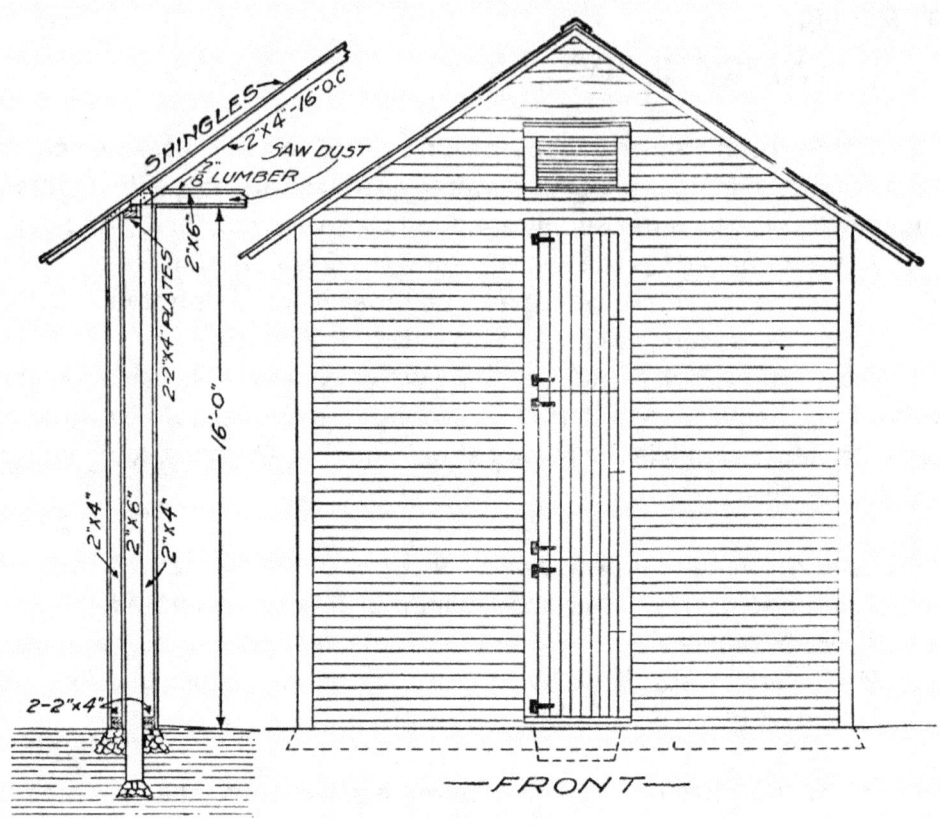

FRONT

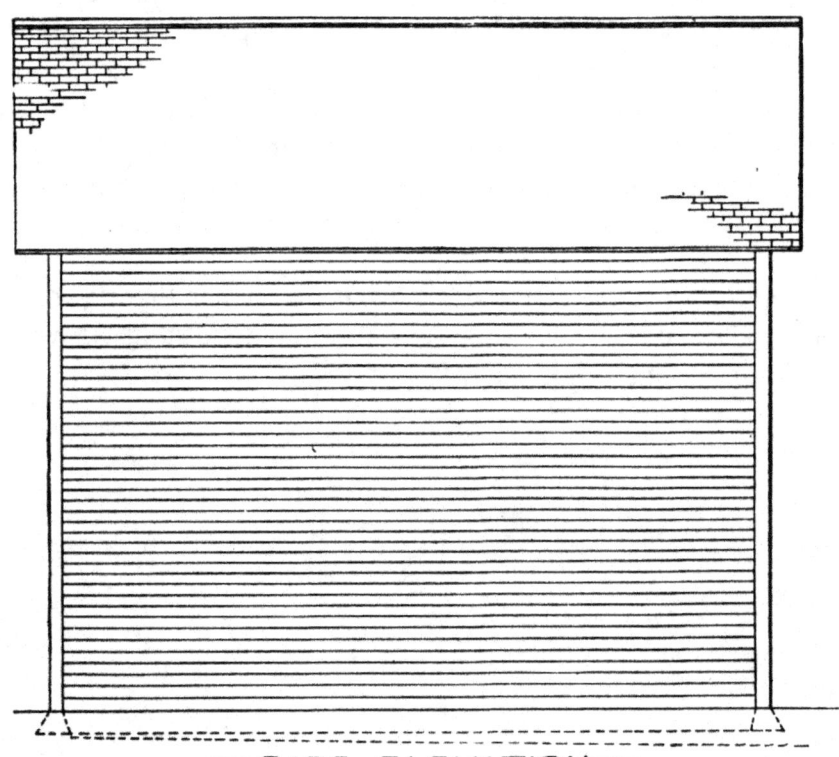

SIDE ELEVATION

# CHAPTER ELEVEN:
## ICE HOUSE

Ice was, and still is, important - with no refrigerators to keep perishables cold, the hot summer months saw a large amount of valuable food be spoiled. Ice is a forgotten winter harvest. Areas of frozen waters were swept and kept clear to help the ice grow thicker. Snow is a natural insulator and left on the farm pond, it creates thin, poor ice for keeping. After a suitable area was cleared, groomed and prepared, and ready for harvesting, one would then saw a block of ice and store it in the ice house for use in the summer months. Ice was best harvested when it was about a foot thick.

The Osborne Almanac, produced by International Harvester, reminded and showed the people once again just the most effective and prosperous way to build one. Ice was viewed as a medicinal use, as it was needed in many ways - just think of all the ways that we rely on a ice box or cold items everyday! By having ice convenient in the rural districts, it helped in all aspects of rural life where there is no stores or doctors easily accessible.

Locating the ice house was important - and the winds and summer heat was important to consider. Placing the building in a hollow or near a bank allowed for the ice melt to run into the water. Too close though meant the flood waters could ruin or containment the ice. Too far away from practical usage in the warmer months made the collecting of ice a unfavorable chore. Ice was harvested two weeks out of the year, usually in January, and the mild inconvenience of packing the ice house made up for the other weeks when the ice was needed.

To build an ice house, the following guide was issued with specific details.  The floor of the house should be porous, the center a trench 30" wide, 12" deep, filled with loose stone up to 3" from the top. Then fill to the top with wood shavings and straw, covering over with loose boards. This creates the perfect drain that will carry off all the water and not let in air. Set the posts in the ground to build on, placing the 6"x6" posts about ten to twelve feet apart.  fill in between with 2x4" studding and set flush with the outside. Put boarding on this. On the boarding, nail 2x4", putting the finally outside boarding on them. This creates an air space of 4."

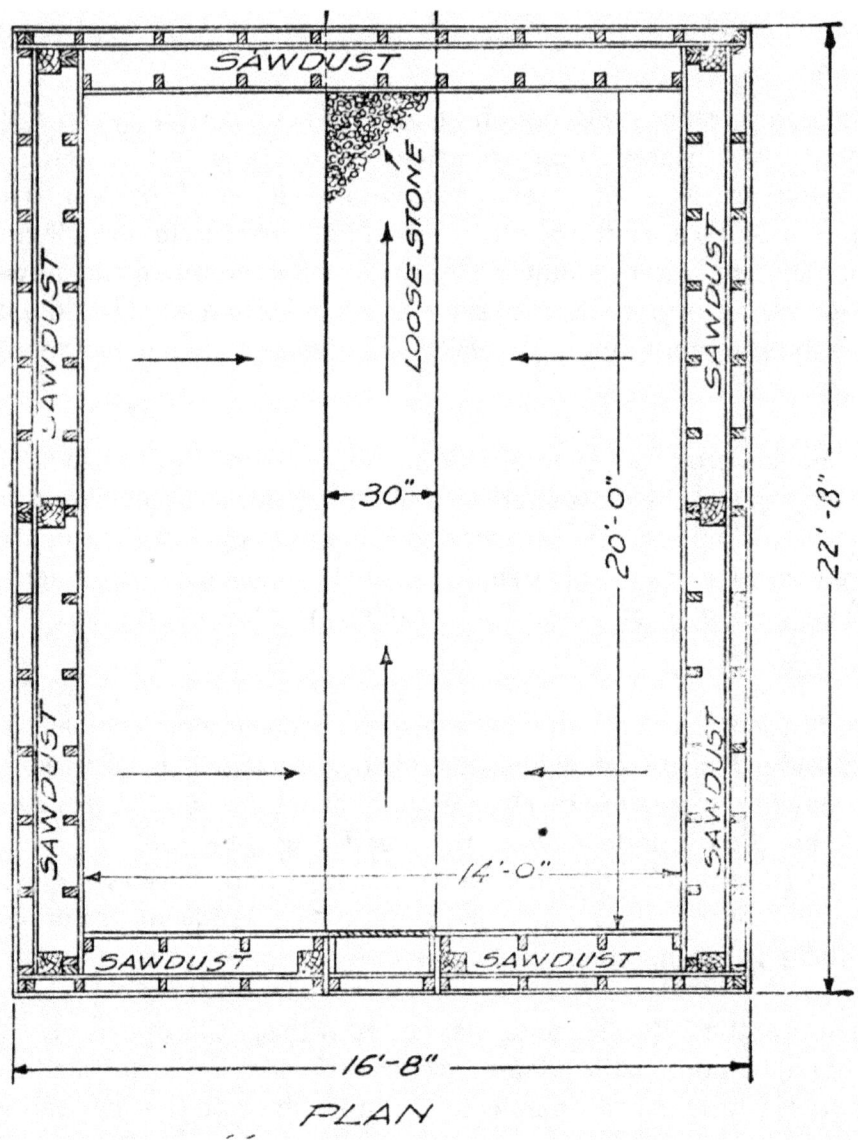

Ice House floor plan, featured in the Osborne Almanac, 1912,
a division of International Harvester
*Tomac Collection*

Inside, set another 2x4" studding on the outside of the 6x6" posts. This creates a 10" gap for filling with sawdust and shavings. Seal on the inside with boarding and most importantly be sure it is filled solid and dry. You now have the body of the ice house, with an air space of 4" and inside 10" of filling, making a perfect house that will not easily warm through. Have the inside smooth so the ice will settle and not catch. Make an ordinary truss roof, which should product at least two feet and can be of board or shingles. The loft inside is to be boarded over as it stops all heat from the roof.

Openings to air should be as few as possible and made to close tight. If a more substantial construction is desired, the foundation may be made of brick or concrete.

Caring for Ice: Once stored, do not neglect it. Make sure the top is always covered. See that you do not get air holes through the sawdust as that lets in the hot air and melts the ice fast. When sawdust can not be secured you can use oat or rye straw.

As to the size and capacity of the houses, with solid and well-frozen ice, it will hold for every foot in height: 14x20, 5 tons; 14x25, 6 tons; 14x30, 7 tons; 20x25, 9 tons; 20x30, 11 tons; 20x40, 15 tons.

The amount to be put up depends, naturally, on the amount and purpose to be used. The average farm requires about 200 pounds daily if ice is used to cool the milk, butter, cooked food, vegetables, and so on.

The number of months ice is used depends on the season, one can be sure that at least 6 months of the year, one will require ice.

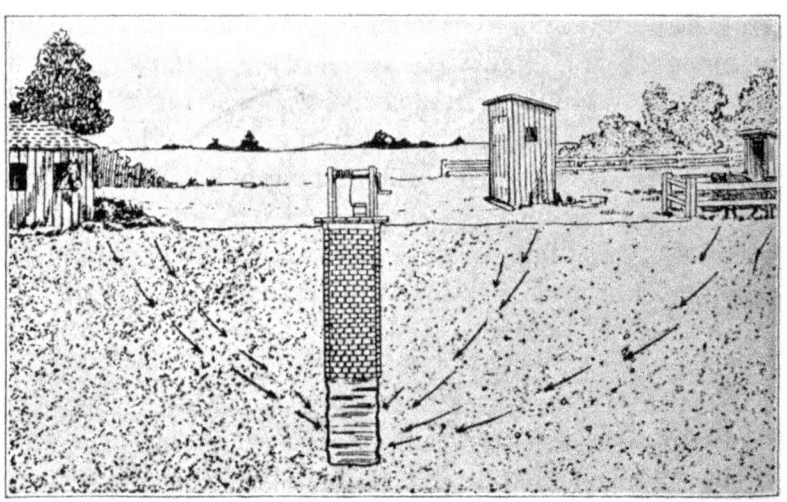

How water becomes polluted in the well. Arrows show course of drainage from the barn yard, manure pile and outhouse

Drawings from International Almanac, 1918, showing the importance of ground water and manure runoff. Installing a septic tank stops the spread of bacteria.
*Tomac Collection*

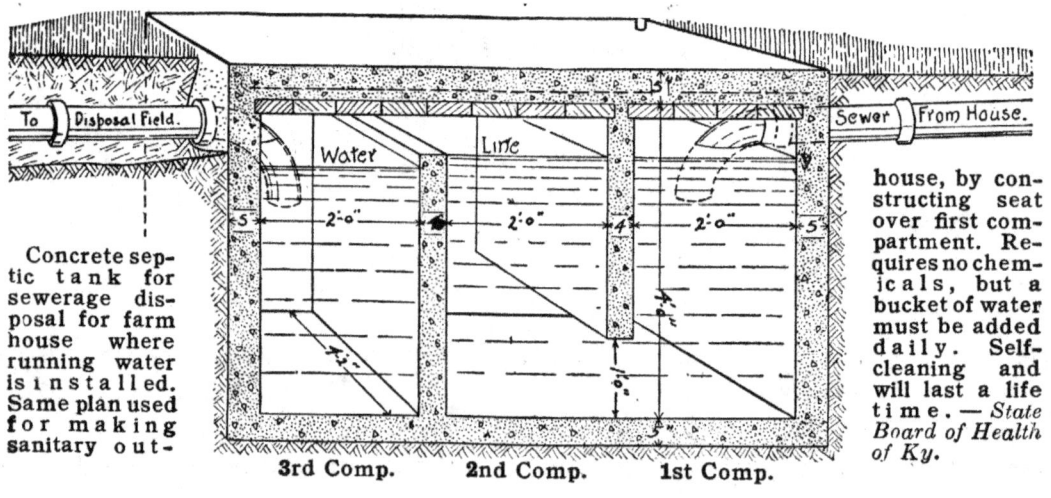

Concrete septic tank for sewerage disposal for farm house where running water is installed. Same plan used for making sanitary outhouse, by constructing seat over first compartment. Requires no chemicals, but a bucket of water must be added daily. Self-cleaning and will last a lifetime. — *State Board of Health of Ky.*

# CHAPTER TWELVE:
## SEPTIC

In the early 1900s, a majority of houses did not have an indoor toilet. New house plans were still being circulated without this idea that you would bathe or use facilities inside a room in the house. It was common for the kitchen (or back porch in the summer) to have a large tub that the family would take turns to bathe in. The privy, or outhouse, was a very active part in country and town life. The outhouse was a small building set over fairly shallow hole, usually not far from the other buildings of everyday use. A basic building, with a solid floor and a bench across the back with a 'seat' or hole in which personal business of nature would be conducted. Many times it also served as a location to drop unusable waste as well, such as broken pottery or glass dishes that would, over time, break down.

The wrong location of the septic could contaminate the ground water, particularly if the commonplace open, hand dug well, was nearby. With a better understanding of the water systems and diseases, often the death rates in the country remained higher than the city when the water supply and water treatment plants were installed to assist the larger populations of people.

Understanding diseases and water contamination from a researched and scientific view in 1918 led to another push for sanitary conditions on the farm. By installing a septic tank on every farm, it brought about a safe and healthy life, no sewer line or cesspool, no smell, no waterlogged black fetid matter seeping into the fresh water well, no organic chemicals, no disease, no germs.

The septic tank became the climax in economy, efficiency and productivity on the farm - once it was installed, there was peace of mind that the water would be safe to drink without fear of becoming sick.

Where Kemp 20th Century Spreaders are Made.
Newark Valley Works, Newark Valley, N. Y.

International Harvester's first headquarters building at 7 Monroe Street on the northwest corner of Michigan Avenue. IHC later sold the building and property to the University Club in 1907. The building was the headquarters of the McCormick Harvesting Machine Company from 1899 to 1902  WHS#8717

Top: Inside Weiss Equipment, Frankenmuth, MI before 1928. Many early IH dealers were also a blacksmith shop, making and repairing implements.(Tomac)

Bottom: A modern, model location, Kercheval of Sheridan, IHC dealership, 1929; WHS #9340

# CHAPTER THIRTEEN:
## DEALERSHIPS/ PLAN AND SERVICE LAYOUT

In the early days before the formation of the International Harvester Company, just about every other store front sold a type of farm implement. There were hundreds of small and not so small names to be found in various locations. Big names like McCormick, Deering, Milwaukee, Allis, Case, or Emerson were found with a store in town or nearby.

Smaller companies were more localized, selling their tractors and implements to local farmers, some successfully like Friday, Knickerbocker, and Olds Tractor Co, (not to be confused with RE Olds Motor co, from the same city). Even my own small town had their own brand of implement to service the area, Big Rock Plow company.

Many of these implement and tractor companies were unable to put the salesmanship, workmanship and business knowledge together to grow or expand. Some met their end when the structure caught on fire. Some could not provide a reliable or consistent machine. All were inventive and unafraid to risk everything for the betterment of their community. The market crash of 1929 put many of these small backyard-like tractor companies in positions to sell to slightly larger companies or to go out of business. Small businesses that were unsuccessful are not to be forgotten or ridiculed, as these men were instrumental in the development of new ideas and concepts - after all Cyrus McCormick himself was one such person. It had taken Cyrus ten years after the first successful demonstration of his reaper before he started to sell the reaper to farmers.

Sales of the reapers and implements, more often than not, came from the salesman's effort in the community. To help a farmer with no immediate direct benefit, simply because the salesman was in the right place at the right time, often meant the farmer, when it was time to purchase a new tool for the farm, would seek out that particular salesmen. Dealers were encouraged to hold meetings, educating farmers and keeping them up to date in the latest agricultural production methods.

The successful tractor and implement dealers knew that in order to stay in business they needed to go to the customer, traveling the back road and country sides where people lived. They also had to stay educated and informed of the newest implements and technologies that the company they worked for had available, as well as know what the competitors offered. They needed to be able to provide a reliable product, service, and education or training on how to use and profit from this machinery to the farmer.

When everything was shipped by railway, the dealership did not need much showroom space, as the structure for sales of the time required the salesman to travel to the farmer with a model, or the real item, showing the farmer the benefits and how to use the implements. The salesman wrote the order up when there was a successful sale, sent it to the Branch House, which was a large storage facility in a much larger town, and when enough orders were gathered, the Branch House sent for a trainload of machinery to be delivered from the Factory. Once the train load of machinery was received at the Branch House, it was sorted and then forwarded on to the dealership in the small town, who would often know days ahead of time when that order would be arriving. Larger dealerships would create a special "delivery day" event in town, line up all the new implements right on main street to show everyone how the International Harvester Line was a prominent part of the community. Farmers would arrive and throughout the day would collect their order and head back to the farm.

*This view is a line up of the No. 3 International New Low for "Spreader delivery day" in Stillwater, Minnesota 1914. WHS#11064*

*WHS#9071*

Being a part of the International Harvester Company, as a dealer in a small town far away from Chicago, Harvester sent out booklets and manuals and all types of information to help the agent become a successful dealer. It was the Dealers responsibility to get this information to the community.

The company was a strong advocate of selling service with the goods. Understanding that the initial sale is what the farmer needed at the time, and the profit to the dealer came with repeat purchases.

There were four service points that the Company stressed to the dealer in 1919 that are still true to this day in any customer service area.

1. Have what the customer wants

2. Have it when he wants it

3. Get it to him

4. See that he makes good use of it and that it serves his purpose.

It is important to have parts on the shelf for the farmer to make repairs, parts that are relevant to the machines in the district. No one wants to drive to town for a part that is not available. For a farmer, this breakdown could cost him more than just the initial outlay of fixing the part. A delay in repairs could mean bad weather and unable to get the crops out of the field the next day, resulting in yield loss. Less money for crops means less money available for new equipment, which leads to the dealer not selling new machines, and the whole circle of prosperity and a successful farmer ends.

In 1919, International Harvester sold small stationary engines as well as a wide array of implements and machinery that was pulled by horses. The tractor was just beginning to be established as the workhorse of the future. Most Harvester dealerships were located in the middle buildings of the main street next to the bank, as the company suggested in the early days of operations.

In 1937, the dealership's early building needs were met, and had the ability to expand inside the existing structures for service and parts. Updates were visual, through painting and reorganizing departments.

Above: Inside a repair shop at a dealership. The brightly painted walls and designated repair areas, or bays, helped to make repairs timely and efficient. WHS#6954

Right: The original storefront in Frankenmuth, MI about 1970, just before relocating to their current location. (Tomac)

The Harvester Company encouraged dealers to have a repair shop available for the more difficult repairs that the farmer could not complete himself. With a full line of equipment available to the farmer, the company stayed in the front of the competition by offering a service that was universally recognized throughout the country, and the dealer was a fine representative of the International Harvester Company. In turn, dealers encouraged farmers to repair their machines as soon as possible, getting their machines ready ahead of the upcoming harvest season, to avoid costly repairs or unavailable parts. A farm repair week was recommended by the dealer to the farmer, as it is important to have equipment repaired or replaced as needed. This still happens today, with customer appreciation day, usually held in the spring, with parts or order discounts offered for that day as well.

The International Harvester Company preferred their dealers to be located on the same street in town as the banks and other major businesses, to promote themselves as a quality company, not in a back street or in a rundown location. The farmer needed to be able to find the best place to purchase implements from and that is just as important as taking care of people's money. The greatest pursuit of man is agriculture. Without agriculture, man has nothing.

The company idea of productivity and prosperity was showcased in the dealership. Presenting the branches in the same manner throughout the country, any farmer that wished to write a letter to a relative, venture to a new location or kept up-to-date with regional happenings, knew that he could expect to receive the same service and attention in any International Harvester location.  Presenting a clean and orderly business is a matter of habit in the shop and a matter of pride on the presentation of the servicemen. The customer naturally formed a positive idea on the dealers' ability to deliver service.

The first major campaign from International Harvester to modernize the dealership in to a uniform and pleasing place of business was shown in the 1937 brochure that was sent out to the dealerships, highlighting some of the recent changes that had taken place in locations across the country.

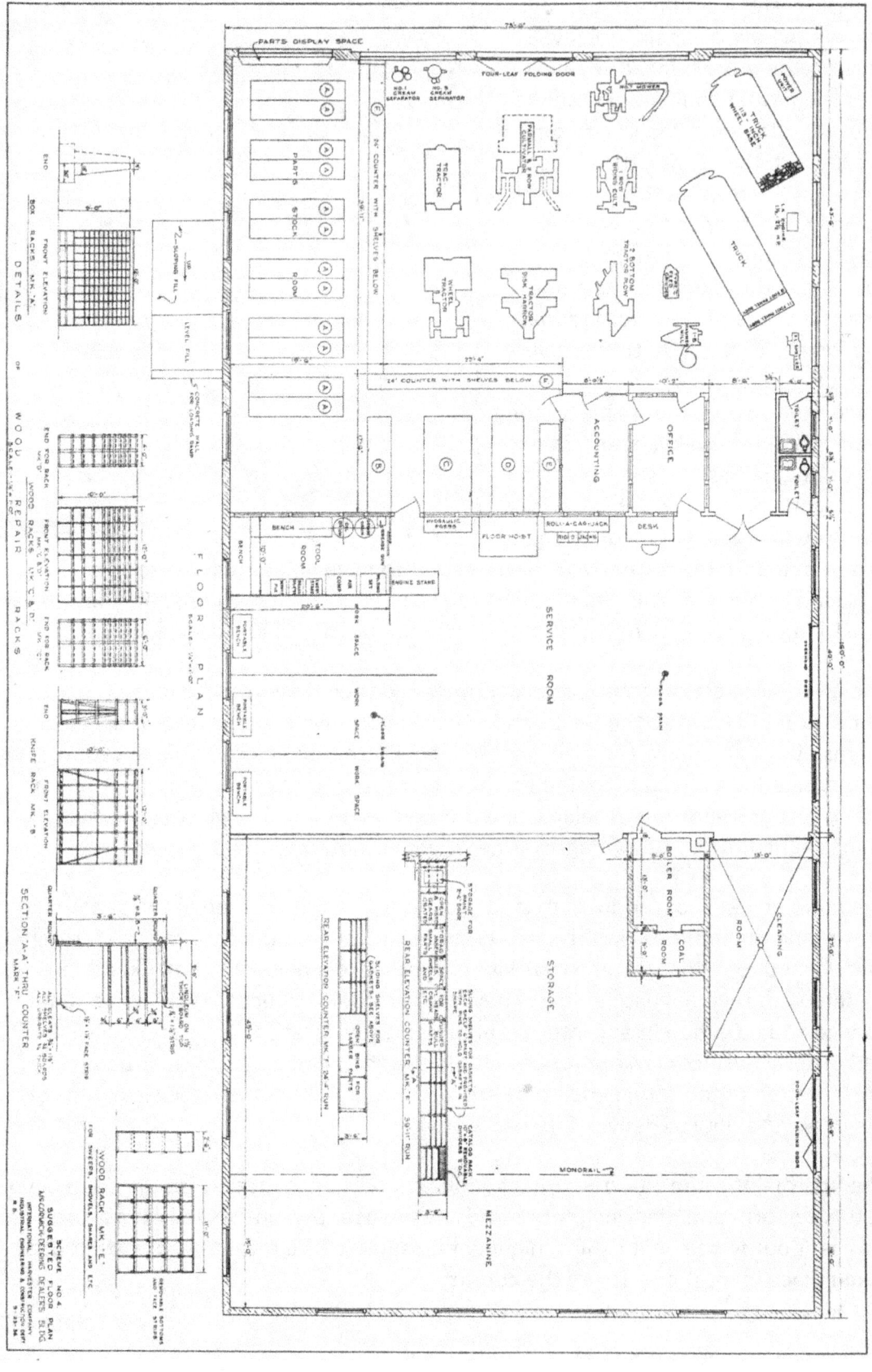

Painting the outside of the building cream, and the inside with Harvester red and blue against a white background showcased the floor samples of equipment from International Harvester. Refreshing the showroom as well as the service, parts, and overall area of the existing building with efficiency and economy was stressed. Three major areas were created, Parts, Service and the Showroom or Sales.

The Parts department was centrally located for both the convenience of the customer and the mechanics. Bins and racks were all made uniform and the same height, locations labeled and parts grouped together. Ends of the parts bays were used for the larger items that could be hung, like gaskets and belts.

Service and repairs were able to be completed faster with just a few simple changes, creating individual repair bays for the mechanic, rolling his workbench and tools from location to location, always within reasonable distance to the larger tools, like a drill press or magneto repair bay.

In 1938, in the Plan and Equipment Manual, Harvester issued these ideas to the dealership in order to encourage the owner to present himself and his employees in the best possible way, right down to the appearance of the floor. "Paint the ceilings and walls, clean all windows and skylights regularly as well as washing the lights and reflectors at regular intervals. Keep all oil cans and miscellaneous items off window sills. Scrape the floors and scrub with a strong lye solution. When the floor is thoroughly cleaned and dry, paint lines on the floor of the service stalls. Portable benches should have tops cleared except for the mechanics tools. Junk boxes for scrap parts are to be kept well out of sight. Good visibility is key to attracting business and increasing profit. Walls and ceilings should be painted in light colors to reflect the most amount of light, and the mechanics will not be subject to light fatigue."

This page, Right:
Interior view of the Vance Henkel Company, Taken from the front corner of the entrance, you can see the customers perspective as they would have entered the store in 1938. There is a separate entrance to the service department from a side street. WHS #147454.
Opposite page, Left: The 1938 service plan layout issued by International Harvester. The Plan and Equipment Manual, covered all aspects of a dealership, from the tools used to the uniforms worn. (Tomac Collection)

This 1938 brochure encourages the dealer to update and modernize the dealership, including layout plans and ideas for an efficient, pleasing and prosperous location.
WHS #147451, #147458, #147452

This dealer 'Base of Operations' is lovingly kept original as a museum by the Pennsylvania IH Chapter 17 and is also their meeting room.
(Tomac collection, photo above by Matt)

Many dealers who participated in these changes were happy to share that their business had increased with the appealing storefront, customers appreciated the view of the machines in the large front display windows as well as the quality of service and repairs were improved.

The next major change to the customer appeal from Harvester was issued as early as 1944 with the still familiar Pylon Building - or Prototype Building. This building was readily recognized as a large red pylon with the IH emblem jutted out of the one-story building. The company referred to this building as its dealers' Base of Operations.

A decree in the result of the 1918 Anti-trust Lawsuit stated that International Harvester was not allowed to have more than one representative in a town selling their products. This meant that there could not be a business for the sale of the harvesting machines and another business selling the implements. The two lines had to be sold from one location. In the post-war distribution, this consent decree was taken into consideration, when developing plans for the expansion of its full line of equipment, tractors, trucks, and industrial machinery. The service layout was already established in previous publications sent out to the dealers, with details of the workbenches, toolboxes and clothing to be worn.

The 1944 Base of operations included display racks, shop facilities, new parts displays, and the distinctive appearance of the building exterior. New buildings to be built needed to be able to house all the upcoming new lines as well as the expanded lines that were already carried.

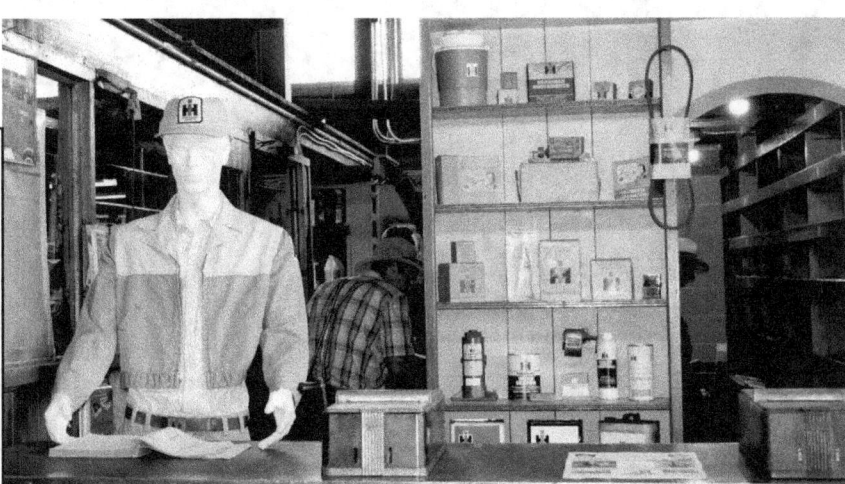

1946 also brought a change in the dress, the distinctive gray uniforms replaced the dark blue attire previously worn. Photo taken at PA Chapter 17 Building. (Tomac)

This view of the Dealer "Base of Operations" in Millville, PA, shows the large front display windows, rear service door, with a matching door on the other side, allowing for drive through repairs. The now familiar pylon adorns the front entrance to the building. The slope of the building site allowed for a unique basement storage feature in this building that belongs to the PA Chapter International Harvester Collectors Club. Below is a view from the inside showroom. *(Tomac)*

After World War II, the model of the salesmen operation changed. The Branch Houses remained as a large warehouse, and the dealership took a more prominent role in having stock on hand to sell to the farmer. The farmer no longer had to wait for the large order to arrive. He could go to the dealership and bring home a new implement or tractor that same day.

The postwar planning for worldwide plant expansion and dealerships by Harvester showed an attempt to improve customer connection with the dealer through better service of the Company's complete line of farm equipment as well as trucks and the newly added line of farm refrigeration.

Harvester issued a manual that set out the plans under this new model of operations stating:

"The dealers base of operations plays a major part in the successful conduct of his business. Because of this - and because we are always willing and ready to assist our dealers in solving their operating problems - a great deal of time, energy, and thought has gone into the preparation of the suggestions contained in this manual."

The Company shift in 1976 in the structure of their business would have allowed for yet another view in the way dealerships were presented to the customer as well as to keep up with the changes in size of the tractors, trucks and implements available. This is evident in the signage, presentation and 'earthy' tones in the uniform dress changes as well.
*Above: The new Weiss Equipment location, Frankenmuth*
*Below: Inside the repair shop at Weiss Equipment at the original location, tight spaces were no longer an issue when the company moved north of town.*

Harvester continues to state the contents of the manual are only suggestions, as there is local considerations and materials to be considered. The Company offered three plan sizes, depending on the volume of sales and service potential. Also covered in the manual were the sales and storage areas, both inside and outside the building, offices, service and shop areas including arrangement, display areas for new machines and parts, office space, customer conveniences and facilities. The locations in respect to main highways entering towns and where that was in relation to the main business centers was also covered. No longer able to house all the new lines as well as the old, businesses were out of room in many downtown locations.

The way that the consent degree was written in 1919 also meant that the dealer who operated the business could not open a branch in the same town, as it was a violation of the operating law. The company's plan in selling this idea to the dealership was left to the various branch managers. A set of blueprint plans were sent out to each branch in May 1945 with instructions that it be turned over to the person designated to handle the building and construction needs.

The plans were highly detailed drawings of a building 81 feet 10 inches by 95 feet 4 inches, outside dimensions, with efficient departmental lay outs, retail space and exterior views. This 7,520 square foot building size was sufficient to handle 6-8 mechanics and gross retail sales up to $125,000.

Two other plans were also available; a 96 foot by 98 foot building of 9,024 square feet, providing for 9-12 mechanics, $225,000 gross retail sales; and a 96 foot by 114 foot building with 10,528 square feet providing for 15 mechanics and $300,000 gross retail sales.

At least two more large manufacturers also had definite plans available and several other smaller farm-related companies mentioned having a better service and sales space created as well. The war was over and all the companies were part of the agricultural boom for a few years.

Left: New tractors welcome customers on opening day at the new Weiss location. (Tomac)

**Good Personal Appearance Increases Service Volume**

In the Country          In Town

The Serviceman is meeting the customer daily, either in the country or in the service station. Impressions created by the Serviceman reflect upon the dealer's entire organization.

A neat, clean appearance of the Serviceman will go a long way towards forming the impression of businesslike efficiency which every good organization should have. Then, too, an orderly and clean Serviceman has more self respect, works with a better spirit, and with considerably more efficiency than the Serviceman who doesn't care. Regulation uniforms aid in good personal appearance.

The following garments are available in Forest Green or Oxford Grey shades:

**Blazer Jackets    Shirts    Laced Breeches    Regular Pants    Caps**

Coveralls are available in Hickory stripe.

Uniforms with the McCormick-Deering International Service Emblem can be secured from any International Harvester Branch.

When ordering garments be sure to specify the sizes as follows:

    Neck size of shirt             Cap size
    Sleeve length of shirt      Jacket size
    Pants waist size              Coverall size
    Pants inseam measurement

---

The standards in business from 1938, bringing uniformity throughout the whole face of International Harvester, began with the appearance of the service and sales men. *(Tomac)*
Below: Group Photo of the Employees at Weiss Equipment, Frankenmuth, Michigan on opening day of their new premises. (Tomac)

*This Page: Samples of the brochure from 1976 that was sent to the dealership promoting the facilities facelift. Below, a closer view of the kit sent to dealerships. (Updike)*

When the 'next generation' of tractors were introduced, it appeared that the current dealerships with their pylon would also, again, undergo changes.

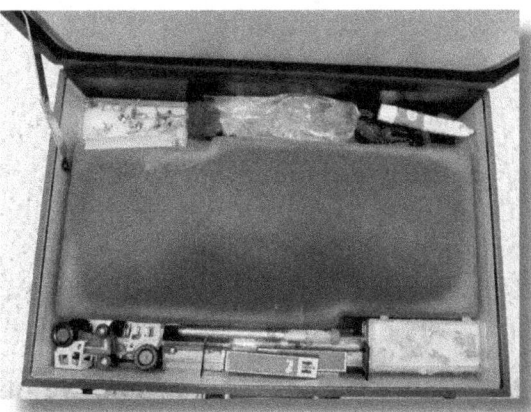

The 1970s were defined by bright earth colors and tones, and this reflects in the Company's encouragement to their dealerships for updates to the existing building - Bronze color metal fascia with gold paint on the exterior walls, finished with natural brown brickwork around the showroom windows. The dealership name in large white letters on the building fascia clearly identified the business. New primary and secondary signs finished the simple update.

The McCormick name drew the customer into the dealership and the customer service kept them there. Women appreciated the attention to agricultural advances in gardening and food preservation that IHCSB shared with everyone. She also gladly accepted household conveniences such as the small stationary engine that would operate a cream separator, a water pump, and later, when the company offered refrigerators, air conditioners, freezers, anything that would help make the farm home more efficient in farm production. It was the wife who would encourage the farmer to buy a new tool or implement, seeing the value in productivity and prosperity.

During the war, women were left to operate the farm, feed the world, and carry on as if there were no rations, concerns or shortages. International Harvester held tractor schools for the women, which may have led to the greater interest in household goods for the farm house, or at least it helped to solidify the need to help the farm woman out of the field as much as in the field.

Today, farming makes up about 1.8% of the US population and of that, in 2017, over 36% are women.

Above: A view of the Tractorettes, the name given to the women who attended classes or schools hosted by International Harvester, who learned how to maintain, repair and operate tractors and equipment during World War II. *WHS #7245*

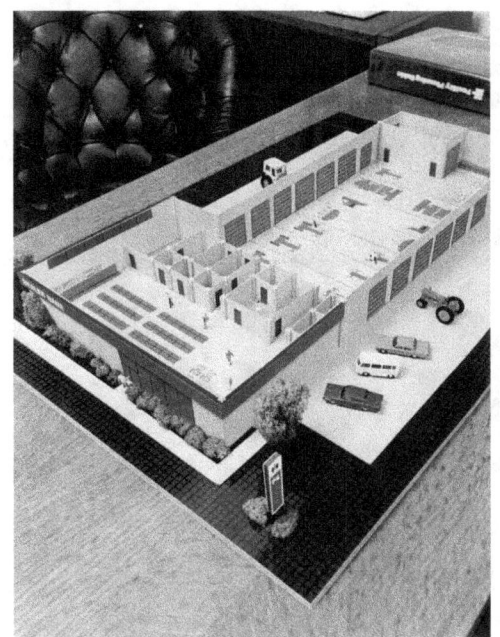

The shift away from the pylon and to the new, sometimes lighted, dealer sign was long and tall. This sign was located closer to the road and the driveway entrance and away from the building.

Marketing from International Harvester to achieve this change in buildings or when considering a new build, was a model kit, presented an attache' case, complete with 1/64th scale tractors, miniature signs, trees, cars and people to replicate visual scale. A planning grid, decals to represent windows and doors as well as service and customer areas were a major part in bringing the dealership into the new era of agricultural efficiency.

Above: A view from the brochure encouraging the dealership to give their business a facelift or build new.
Below: A two page spread from the brochure showing the updated signage and overall facelift. *(Updike)*

The company structure and organization had changed significantly with the introduction of the third major company structure change since 1902, and this may have led to the beginning of the end of International Harvester as it once was known.

Above: A view of the assembly line at *WHS# 6744*
Below: Cyrus McCormick Jr, drives the 200,000th tractor off the line, a 10-20, June 4, 1930 at the Tractor Works. *WHS# 7241*

# CHAPTER FOURTEEN:
## FACTORIES or WORKS

The multitude of International Harvester machinery factories around the world had a few things in common besides the IH name. They were all built to specifications that were issued from the head office. Saw tooth style, with ample lighting and ventilation, paint colors, right down to which direction the building should face. "The high side of the saw tooth will have windows on the north, running east to west". Unfortunately, for the buildings in the Southern Hemisphere, this meant warmer and direct sunlight, instead of the indirect light that was intended and successful in the Northern Hemisphere. In the beginning the factories all needed access to good roads, railways and waters deep enough for shipping. Another major concern is the ability to have employees available for work. Availability of parts, components, materials and other smaller supporting businesses also needed to be in the surrounding area to help with the large scale production of parts, implements and machines.

Most factories today still look for these main principles.

In the early days of 1848, Chicago was the largest most western town, with a population of about 17,000 people, situated on the shoreline of Lake Michigan which overlooked the quickly expanding wheat fields to the west. Cyrus needed to build a factory for his reaper near where the majority of the reapers would be sold, as transportation was not an easy task without railroads. Chicago had yet to become a hub of railway lines when he chose the location for the factory. Despite setbacks in the early days, the McCormick Reaper works produced thousands of machines in the first years. The early years in Chicago saw the establishment of the Chicago Board of Trade, railway lines, and industry. The little town fast became the place to be.

Over the next few years, Cyrus would leave the daily operations to his brother while he sought to expand the reaper business overseas. The exposition of 1851 in England by Queen Victoria and Prince Albert gave him worldwide recognition. McCormick reapers were shipped and sold around the world. Australia received their first reaper as early as 1852. The McCormick Company was established in Australia in 1884 to import and sell reapers throughout the Australian country, based in Spotswood, Victoria. This factory remained as the main location when International Harvester Company was formed in 1902, and when it was established as a separate company in 1912, as a result of the outcome of the antitrust lawsuit.

Above: A newspaper advertisement with an aerial view of the Geelong Works  (Tomac)

Below: Port Melbourne Works, which was home to the construction division in 1957. (Tomac)

International Harvester Company of America became International Harvester Company of Australia, and the only that produced everything in one location. Beginning as a import -style company based in Spotswood, Harvester quickly outgrew that location and saw a need to produce trucks, implements as well as trucks for Australia to export to nearby countries.  The Geelong Factory opened in 1939 with 216 people, the largest factory outside of a major city in Australia. The creation of jobs to the small rural town of 38,000 people  created a stir.   There was a large workforce available to hire and train as needed, much like when Cyrus McCormick built his first reaper factory in Chicago.

 In 1948, the size of the Factory was doubled. Production was increasing and by 1960, the works employed 2,400 people. Finding and deciding on the site with the ability to foresee future growth was a risk that paid off. Geelong was the most diverse factory in the International Harvester world, producing everything needed in one place. Equipped with a foundry, smelter, machines of every type and size, the 45 acre location produced tractors, engines, implements, and truck parts until it closed 40 years later.  It remains the most diversified factory in IH history. Dandenong Works was built in two years and produced the first truck off the line in 1952.  Port Melbourne Works was acquired in 1957 and established as the construction division. Both of these factories received parts from Geelong to complete their respective machines.

In 1961 the Australian branch of International Harvester added two more locations, the Product Engineering Center and the Proving Tract (or Grounds)

Above: A view to the entrance where trucks were manufactured, Dandenong Works.  (Tomac)

Above: The entrance to the Anglesea Proving grounds is protected by a simple gate.
Below: 2,556 acres being developed by machines that were built and under testing during construction. (Tomac)

# CHAPTER FIFTEEN:
## TESTING GROUNDS AND PROVING TRACT: ANGLESEA

The only place in the International Harvester Company history where the machines tested were part of building the facility is at Anglesea, south of Geelong, Australia.

Construction equipment, trucks, tractors, implements all had a part in building the area and were tested at the same time. The nature of the country, a rolling heavily wooded area prone to wildfires, a 66ft wide fire break initially had to be cleaned on the property's limits and 11 miles of fencing erected to enclose the area, all before any building started on the workshops, amenities block or the close-security area.

The first stage of construction to this 2,556 acre site was a 2.5 mile, 24ft wide road, which formed the main motor truck testing circuit and a 2,000ft long sealed loop road used for tractor testing. The second stage involved construction of a seven-mile road test circuit, a main water supply dam and a 300 acre pilot farm. Testing of the company's tractors and farm equipment was carried out on this little location.

Built into the second class road is a one mile section of a 6 percent gradient, recommended by road companies, for testing road conditions. The water supply dam consisted of an earth filled wall about 25 ft high and involving some 50,000 cubic yards of material and was used for the domestic, fire-fighting and construction purposes.

Re-grading and sealing of the main test circuit, construction of a Belgium block (used to create intense vibration on most cab mountings, sheet metal and other components in trucks) and final work on other testing facilities formed the basis of the third stage.

Right: A view from the inside of the workshop and office area at the Testing grounds, you can see a worker writing his reports ans the various stages of investigation on a truck in the background. (Tomac)

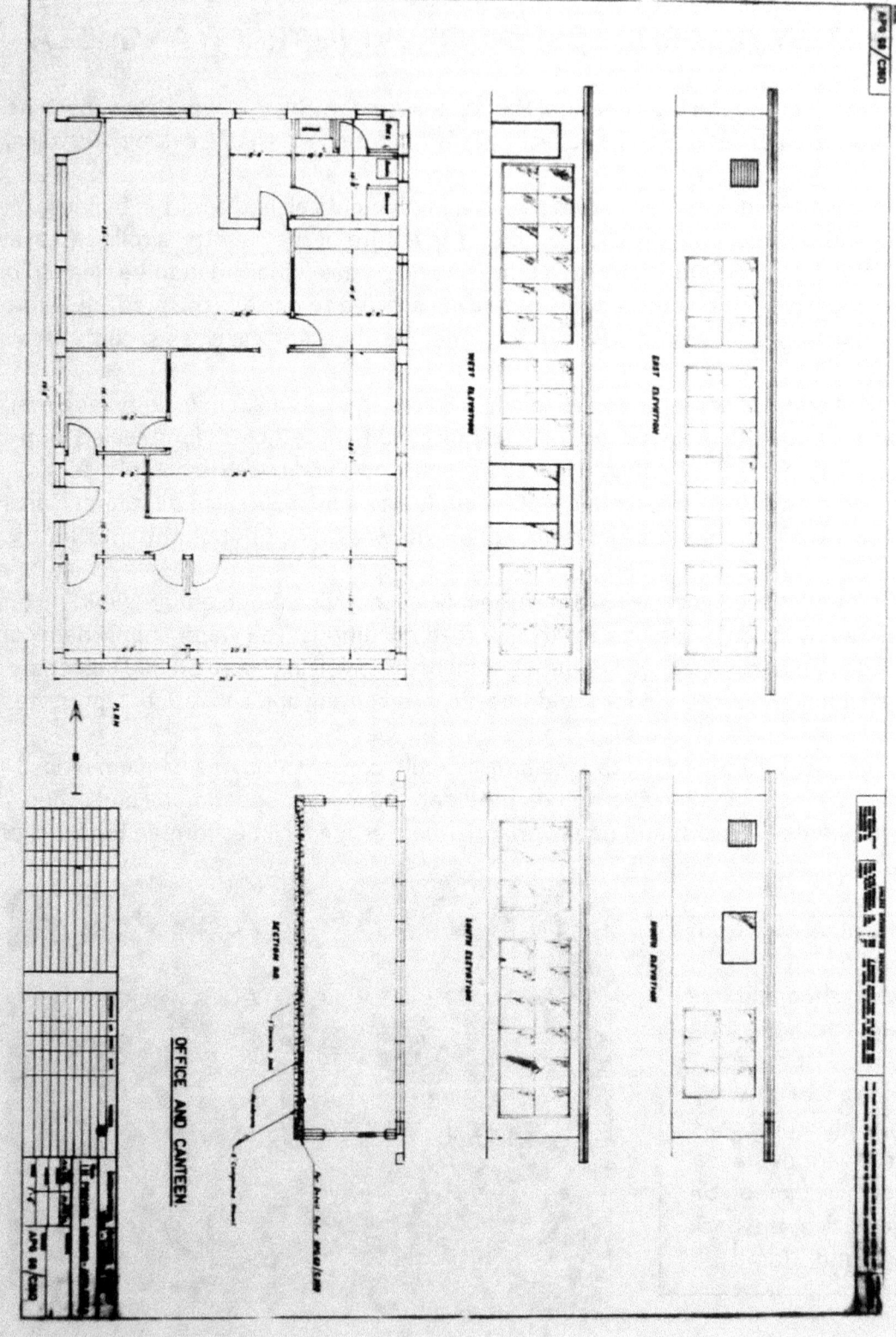

Mud baths, steep grades and rough tracks built to test any vehicle were among the many features at this location. To keep certain mud surfaces in a slippery condition for vehicles and under body components being tested for waterproofing, water frequently was poured onto the mud patches. This was done by means of aqueducts from small dams located on the high side of the road.

A special feature of the proving grounds was the tractor test circuit. Here a driver-less prototype, trailing a load car equipped with instruments to record temperature, pressure and other features, could follow effective electrified wire for 24 hours a day around the circuit. To keep the tractors front wheels on course, the tape initiates a correcting impulse immediately when the wheels start to stray.

Work was carried out in two shifts five days a week, testing and developing the very best in design and engineering.

Left: The Office and Canteen floorplan with elevation views (Tomac)
Below: A view of the Security area on the grounds. (Tomac)

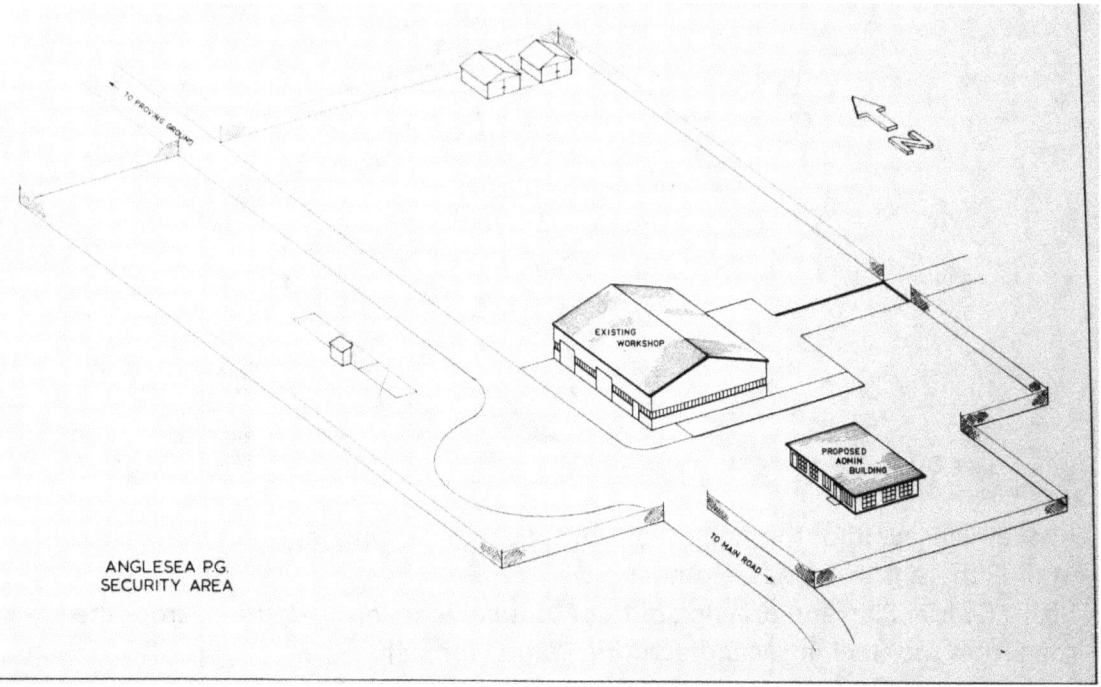

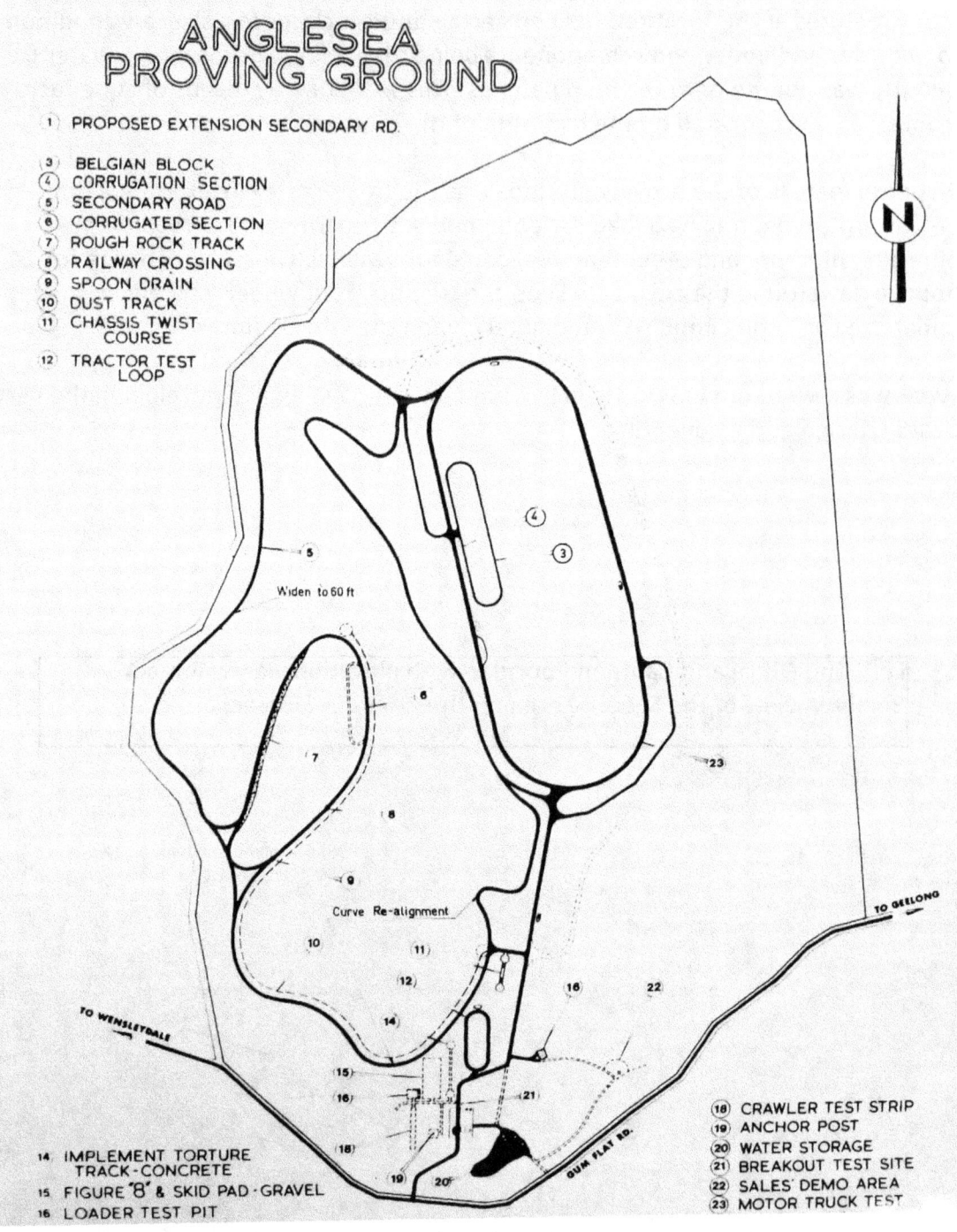

Above: The layout of the Anglesea Proving Grounds, with labels of the locations of the different test areas. (Tomac)
Right: Official Plan and drawing of the Product Engineering Center, across the road from the Geelong Manufacturing Plant. (Tomac)

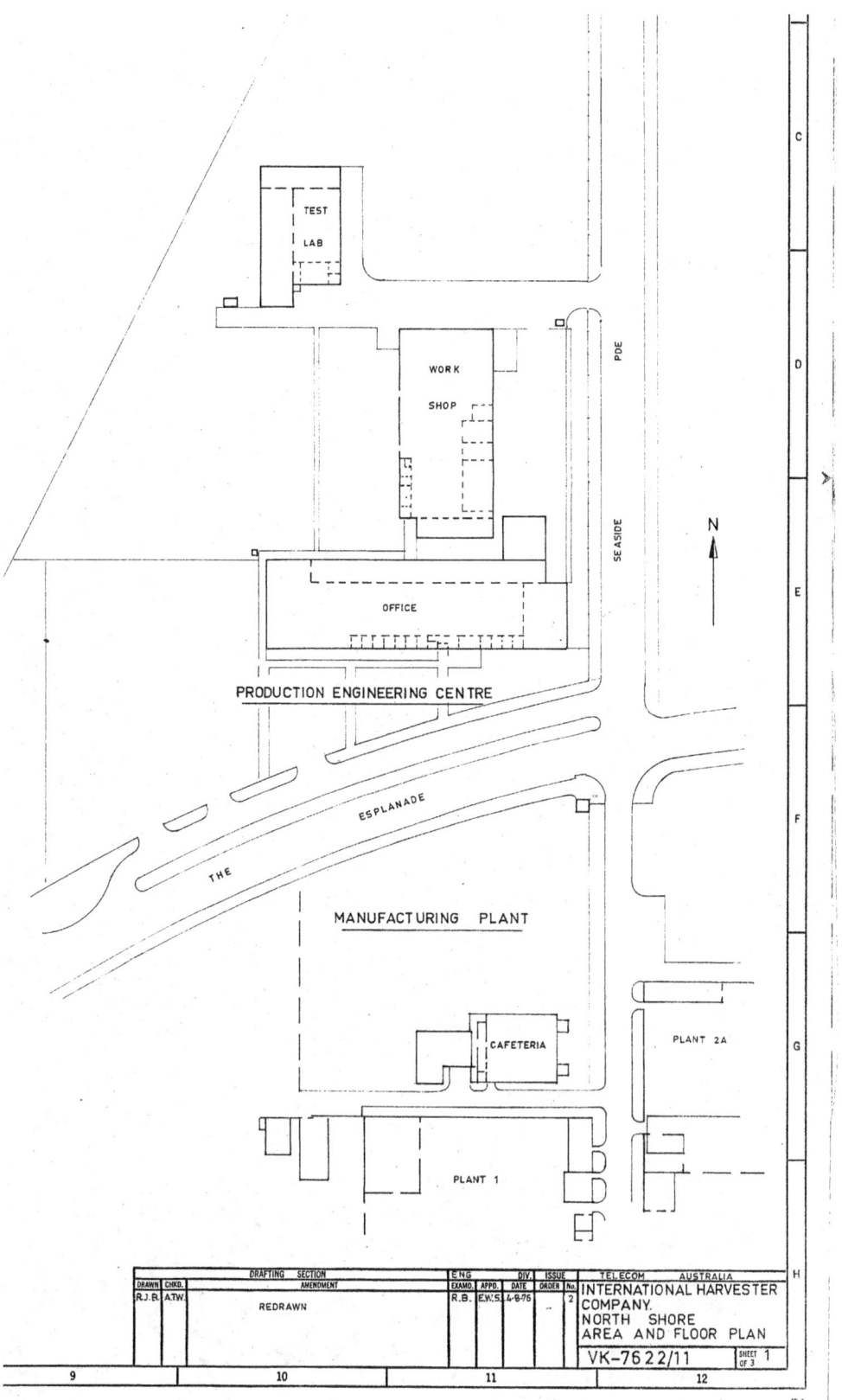

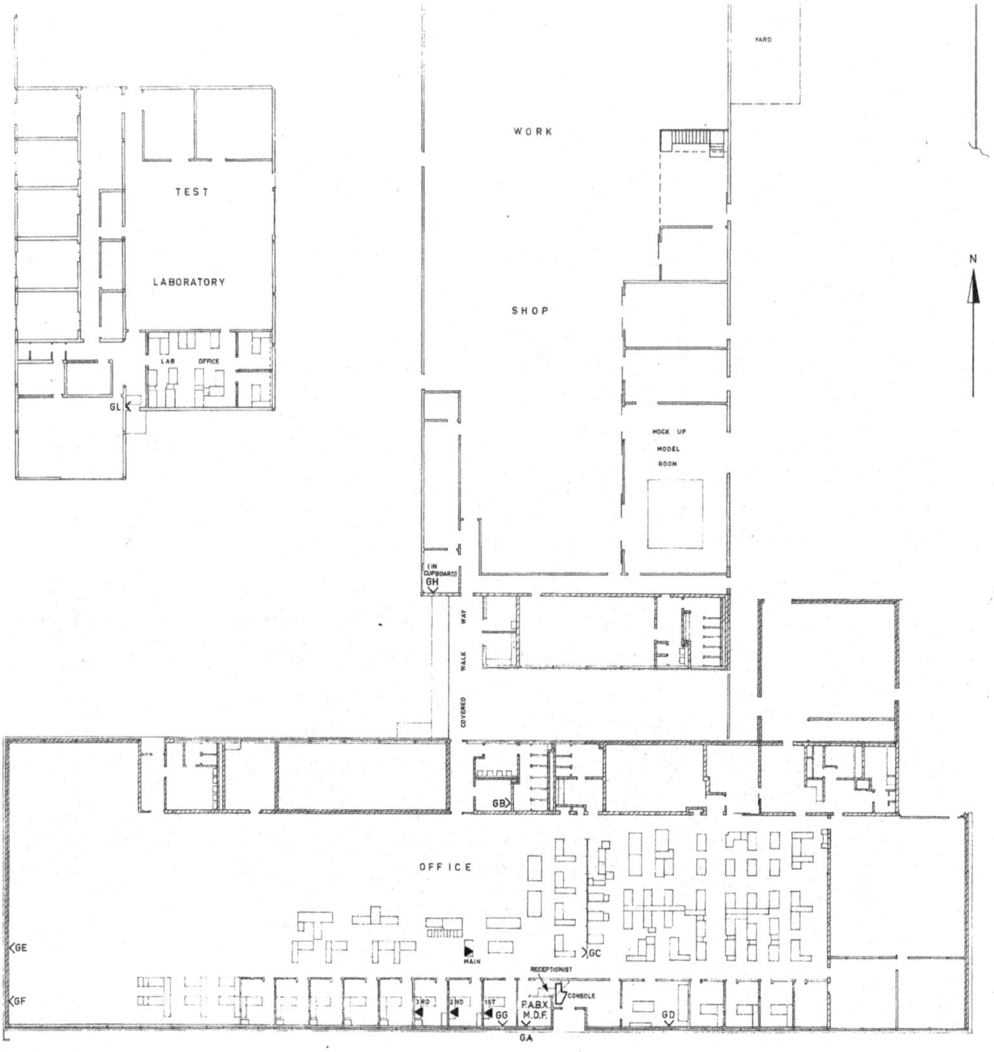

PRODUCTION ENGINEERING CENTRE

# CHAPTER SIXTEEN:
## ENGINEERING DEPARTMENT: AUSTRALIA

Built in 1961 to centralize the entire Engineering team, add 100 employees, and bring the technology up to date for every department, the Geelong Product Engineering Center was located across the road from the Geelong Factory. It was equipped with specialized electronic testing machines to accelerate component testing of fatigue stress, working in hours, instead of week-long field tests.

The Engineering departments were a myriad of activity - everything from drawing to testing to photographing as well as creating life-size clay models. Geelong was a unique center as all of the engineering and development happened in one location. Trucks, Tractors, implements, construction, motors, all had a place under one building.

Everything was drawn, miniature models were made, and life size clay models were created. Photos were taken, tests were conducted, photos developed, repairs and adjustments recorded before it was sent out to the testing track about 30 miles away in Anglesea.

Above: A view during the construction process of the Engineering Center. The Geelong Works can be seen in the background, across the road.
Left, Top: A detailed drawing of the Product Engineering Center layout. The front space was filled with drafting tables and offices, opening to the photographic room, dyno room, testing and model rooms with workshop in the rear.
Left: a view of the workshop area at the Testing Center. (Tomac)

International Harvester Company Wagon Works No. 1, (Weber Works, Chicago, Illinois) where the Weber, "King of All," and the Columbus Farm Wagons and Farm Trucks are made.

International Harvester Almanac, 1909
Weber Works, Chicago, IL
Weber Wagons and Farm Trucks (platform carrier on wheels)

Engineering departments around the world in 1975 were located in the following Cities and Country:

United States:

    East Moline Engineering: East Moline, Illinois

    Fort Wayne Engineering: Fort Wayne, Indiana (Truck Group)

    Hinsdale Engineering: Hinsdale, Illinois (Agricultural Equipment)

    Libertyville Engineering: Libertyville, Illinois

    Melrose Park Engineering: Melrose Park, Illinois

    Memphis Engineering: Memphis, Tennessee

    San Diego Engineering: San Diego, California (Solar)

    West Pullman Engineering: Chicago, Illinois

    Woodfield Engineering: Schaumburg, Illinois (Industrial)

Canada:

    Chatham Engineering: Chatham, Ontario

    Hamilton Engineering: Hamilton, Ontario

Australia:

    Geelong Engineering: Geelong, Victoria

New Zealand:

    Christchurch Engineering: Christchurch

England:

    Doncaster Engineering: Doncaster, Yorkshire

Above: Dandenong Works, with lineup of ACCO trucks ready for delivery.
Below: Drawing of the Port Melbourne Works, construction division
Below Right: Product Engineering Center, Geelong.
All part of the Australia division, Tomac Collection.

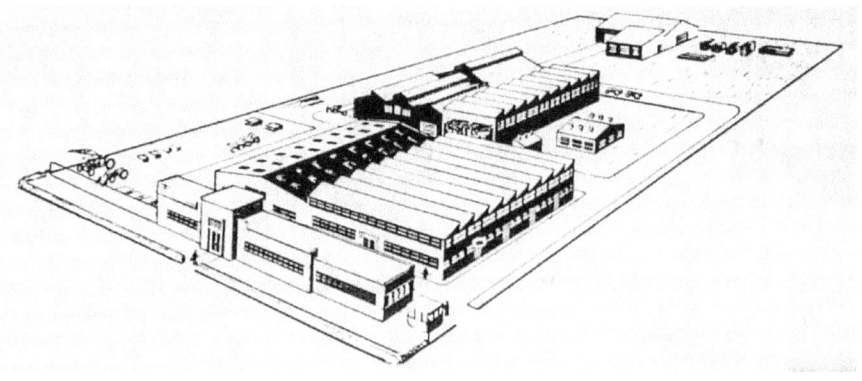

France:

    Croix Engineering: Croix

    Genas Engineering: Genas

Germany:

    Neuss Engineering: Neuss

Japan:

    Kawagoe Engineering: Kawagoe City (Komatsu)

Philippines:

    Manila Engineering: Manila (Macleod)

South Africa:

    Pietermaritzburg Engineering: Scottsville, Natal

Mexico:

    Saltillo Engineering: Saltillo, Coahuila

Above: The official opening of the Port Melbourne Works.

Right: The layout of Port Melbourne Works. (Tomac)

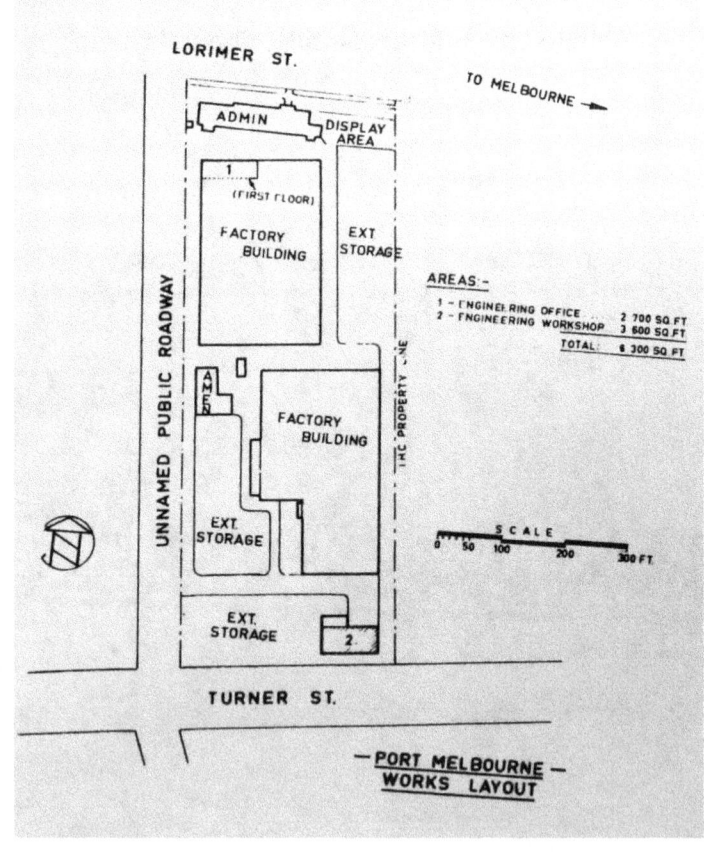

Plants and Factories in use in 1980:

United States
    Canton, Illinois – Factory: Farm Tillage Tools
    Chicago, Illinois – Factory: Components (West Pullman Works)
    East Moline, Illinois – Factory: Combines, Planters,
                                Control Centers
    Libertyville, Illinois – Factory: Front End Loaders,
                                Scrapers, & Off Road Trucks
    Melrose Park, Illinois – Factory: Crawler Tractors, Engines
    Rock Island, Illinois – Factory: Farm Tractors (Farmall Works)
    Fort Wayne, Indiana – Factory: Trucks, Scouts
    Indianapolis, Indiana – Factory: Gas & Diesel Engines, Foundry
    Gulfport, Mississippi – Factory: Construction Equipment
    Louisville, Kentucky – Factory: Outdoor Power Products,
                                Components, Transmissions, Foundry
    Memphis, Tennessee – Factory: Cotton Pickers, Plows, Hay Tools,
                                Nodular Foundry
    San Diego, California – Factory: Solar Turbines
                          Factory: Solar Turbine Drive Packages
                          Factory: Solar Turbine Components
    Columbus, Ohio – Factory: Plastic Components for Trucks,
                              Plastics & Remanufacturing Facility
    Shadyside, Ohio – Factory: Sheet Metal Stampings for Trucks
    Springfield, Ohio – Factory: Trucks, Truck Bodies
    Wagoner, Oklahoma – Factory: Heavy Trucks
    Waukesha, Wisconsin – Foundry: Malleable & Nodular Castings

Australia
    Dandenong, Victoria – Factory: Medium & Heavy Trucks
    Geelong, Victoria – Factory: Farm Tractors, Combines,
                                Implements, Engines, Foundry

Works at Akron Ohio, where International Auto Buggies are made, Harvester Almanac, 1909 (Tomac)

Above: Entrance into the Proving Grounds, the Workshop can be viewed through the trees.
Below: A view of the lighting that adorned the top of the main building at Dandenong works, with inset of the view during the day. (Tomac)

New Zealand
    Christchurch – Factory: Farm Tractors, Implements, Trucks, Wheel Loaders

Canada
    Candiac, Quebec – Factory: Wheel Loaders & Loggers
    Chatham, Ontario – Factory: Trucks
    Hamilton, Ontario – Factory: Farm Implements

England
    Bradford, Yorkshire – Factory: Farm Tractors, Industrial Tractors
    Doncaster, Yorkshire – Factory: Transmissions, Crawler Tractors, Foundry
    Factory: Farm Tractors, Balers

France
    Angers – Factory: Combines
    Chauffailles – Factory: Excavators
    Croix – Factory: Combines, Farm Implements
    Genas – Factory: Excavators
    St. Dizier – Factory: Farm Tractors, Transmissions, Foundry

Germany
    Heidelberg – Factory: Front End Loaders
    Neuss – Factory: Farm Tractors, Diesel Engines, Foundry

Mexico
    Saltillo – Factory: Trucks, Farm Tractors, Farm Implements, Engines, Foundry
    Mexico City, D.F. – Factory: Excavators, Small Crawlers

Philippines
    Pasig – Factory: Trucks, Farm Tractors, Farm Implements

South Africa
    Pietermaritzburg – Factory: Trucks, Farm Tractors

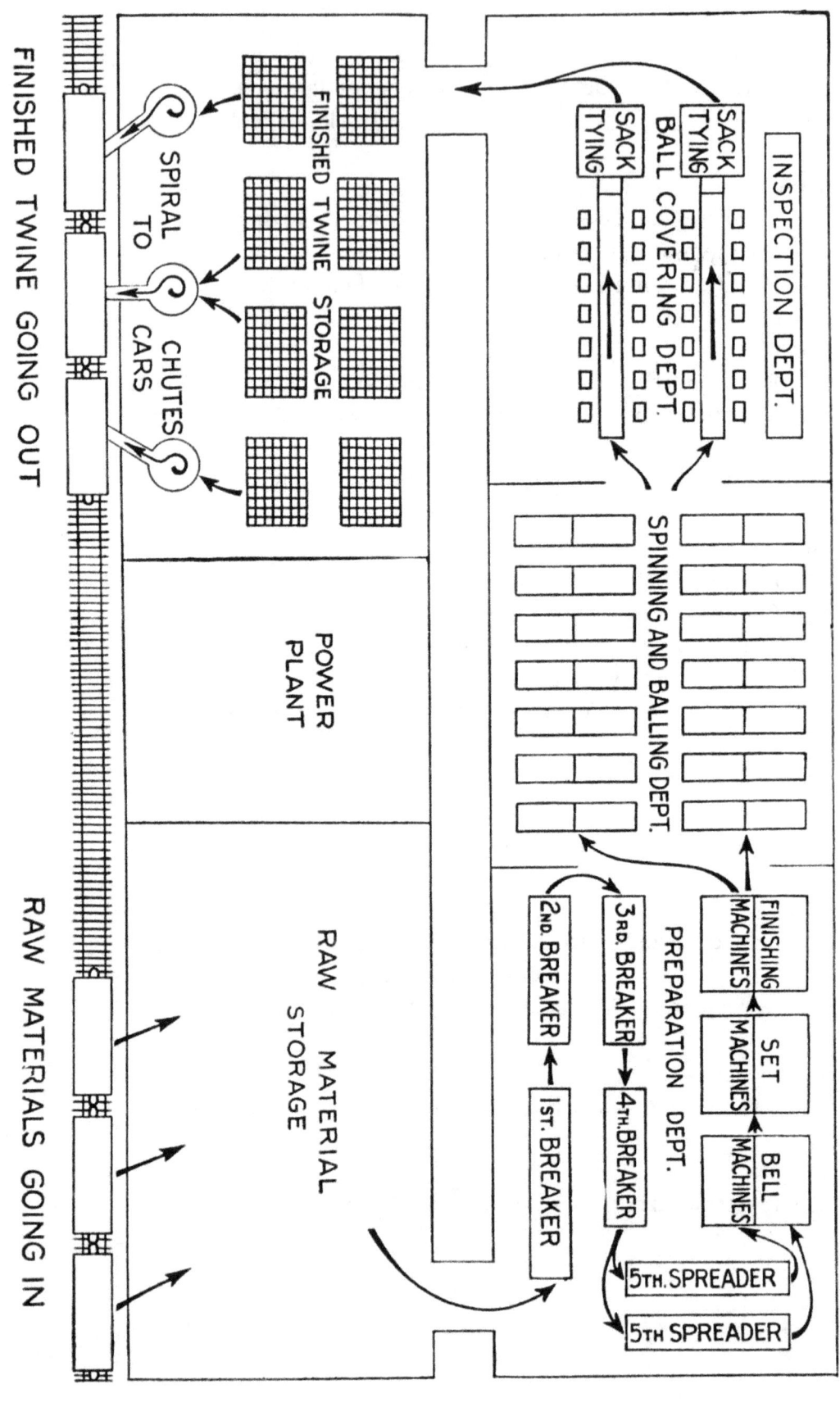

# CHAPTER SEVENTEEN:
## TWINE MILLS

International Harvester needed a fiber to hold the bundles of wheat together, twine. With the development of the twine hold the grain bundles, it allowed farmers everywhere to quickly and efficiently harvest the thousands and thousands of acres of wheat throughout the world. Many years of experimenting were spent leading up to the correct blend of sisal and manila fibers. The failure of being able to use grass, hemp, flax, straw and paper, meant the fiber for the successful blend could not be grown in the United States or Canada.

Mexico, Philippine Islands, Cuba, Java, and East Africa are the main countries where the henequen and abaca plants are grown in the tropical climate. Growing plants on a limestone and coral formation is not suitable soil for grain. The blend of this soil and the warm humid climate is suitable for growing the sisal needed for twine. Yucatan, a small area on the Gulf of Mexico, produced a large proportion of the sisal fiber needed for the twine production that International Harvester needed. The principle source of manila comes from the Philippine islands, and is grown mostly on volcanic ash soil. From the many tropical points in the world, the raw fibers were harvested, baled and shipped to twine mills around the world.

Twine mills were located in the following countries-
United States: Chicago, Illinois; and New Orleans, Louisiana

Canada: Hamilton, Ontario

Europe:
Norrkoping, Sweden;
Croix, France;
Neuss, Germany

Left: The floor plan shown illustrates the general movement of the production of twine, regardless of the building it is in. From "The Story of Twine" (Tomac)

Above: A majority of the fiber for twine is in the Yucatan and the Philippines. This oxen team is bringing a load of fiber to the processing plant. WHS#8084

Below: The home of the Twine facilities in Philippines, the main offices are next door to the facility. (*The Story of Twine, Tomac*)

Fiber has to undergo three processes to become twine. First, the preparation of the fiber. Second, the spinning of the twine. Third and final, the balling and stacking. The large bale that started as a flow of fiber several inches in diameter and weighing more than half a pound to the foot, ends as a strand of twine one twelfth of an inch and more than 500 feet to a pound. WHS#7675

Above: The attention to detail can be seen below a welcoming window, with the Initials of White Deer Lake inset.

Below: Close up view of the side of a cabin shows the attention to high quality craftsmanship.

Photos and drawings are courtesy of US Department of Agriculture and Forestry Service.

# CHAPTER EIGHTEEN:
## MCCORMICK PRIVATE CAMP

The McCormick family was not all business, and they held lands and property in various different places throughout the United States. Some of these places are better known, like California and New York, there are also lesser known places of residence. Mrs. McCormick was known to have houses in multiple places around Chicago, residing in her later years in a quieter location, fondly named the House in the Woods.

Cyrus McCormick (Jr), as all young men of nature are predisposed to, found a respite from his grueling activities as leader of the growing McCormick estate in the woods and hideaways north of his Chicago home. A good friend of the family, Dr. William Gray, from his Princeton days, introduced young Cyrus to the wilds of Wisconsin at an aptly named Island Lake Camp. This camp was a respite in the years following the death of Cyrus, the inventor, as Nettie was 25 years younger than her husband.

Island Lake Camp, now known as Potawatomi, was established in 1887, in Bayfield County, Wisconsin. It is part of the Eau Clair Lakes Chain in northwestern Wisconsin, and listed on the National register of historic places. The simple structures were built by local lumberjacks well known for their skills, Allen Woodruff and George Wilson.

The first structure built in 1887 was the main dormitory, a unique design not found in the area. A dog-trot design, which was two cabins connected by a center court area, all under one roof line. Between 1888 and 1890, the sleeping rooms, dining cabin, ice house, library and one of the double cabins were built. The second cabin and the Wanigan were finished in 1895. In 1888 the caretaker, William Green and Lewis Ramsell, farm hand and horseman for the camp, both accomplished builders, built a boathouse to surprise the Grays.

Dr Gray frequented the Island Lake area until his death in 1901, where a total of seven log cabins had been constructed. The McCormick's replaced the boathouse about 1915, and the cabins are still in use today.

Above: The view from the shore of the mainland to the Whit Deer Lake Island retreat. An impressive sight surrounded by nature.
Below: Plot plan of the Mainland Buildings. *These photos and drawings are courtesy of US Department of Agriculture and Forestry Service.*

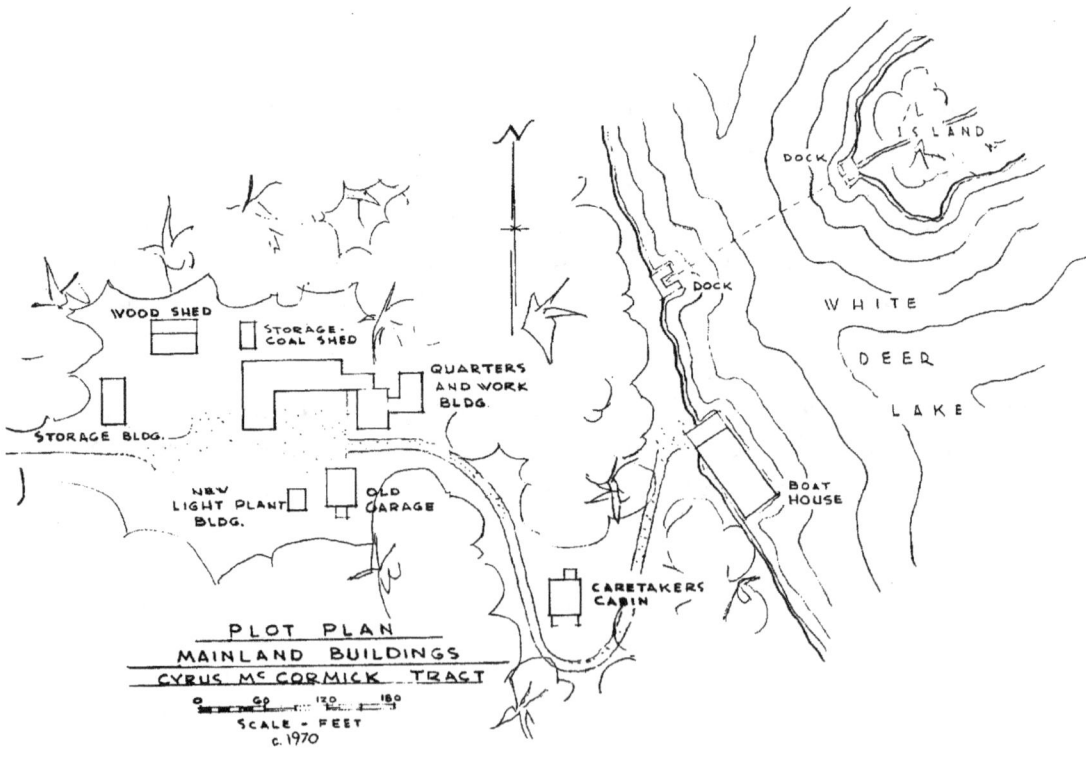

Perhaps it is because of the passing of Dr Gray, a very close friend of Cyrus H, which led him to pursue an even more remote area of wilderness with his close friend and legal counsel, Cyrus Bentley.

Bentley and McCormick were avid outdoorsman much in the way that Gray was. Gray's permanent camp, Island Lake, hosted many visitors over the years. Gray would go camping in multiple locations, selecting a place where two or three headwaters formed, with many small lakes to explore, as he was an avid hiker.

One such location was found in Michigan, and Cyrus was invited to go along with Gray in 1884 to this remote location. In order to reach the camp, the men followed an Indian guide, north along an old trail that began in Champion, MI, along the Peshekee river.

Arriving at the campsite in 1884, Cyrus was astonished to find that Gray had set up large tents on the shore of this nameless lake, which held some of Cyrus' belongings. Gray, as a surprise, had contacted Mrs. McCormick and requested assistance in arranging this camping delight. During the twenty years between the first camp and purchase of the property, McCormick and Bentley often visited the remote area, enjoying the wild game and surrounding scenic beauty and remoteness, a very much opposite and clearly relaxing relief from their busy and social daily family business.

Cyrus Bentley eventually would build a cabin on the shores of the Superior, near the Huron Mountains Shooting and Fishing Club. Bentley, who was also an avid hiker, wanted to build a trail from Fortress Lake, as the two Cyrus' called the area, to the Huron Mountain Club, 26 miles away. This perhaps was the final push for Cyrus to pursue the purchase of the property in 1902.

To give a clear idea of how the International Harvester Company was led in the days under Cyrus H McCormick's leadership, along with the influence of his upbringing to mind his financial matters, it is important to look at the way the property in the remote Michigan wilderness was acquired.

Edwin McLean, a scout who evaluated the area for McCormick, felt the area could be purchased for about $5/ acre. McCormick, well aware that his name alone would cause the property price to increase, asked an agent in Marquette, W.E. Lewis, to act on his behalf. When Mr. Lewis approached the land owner, he was firmly told a purchase price of $10/ acre. John M Longyear, the owner, had discovered that McCormick was interested in the property and did not want to do business through Mr. Lewis.

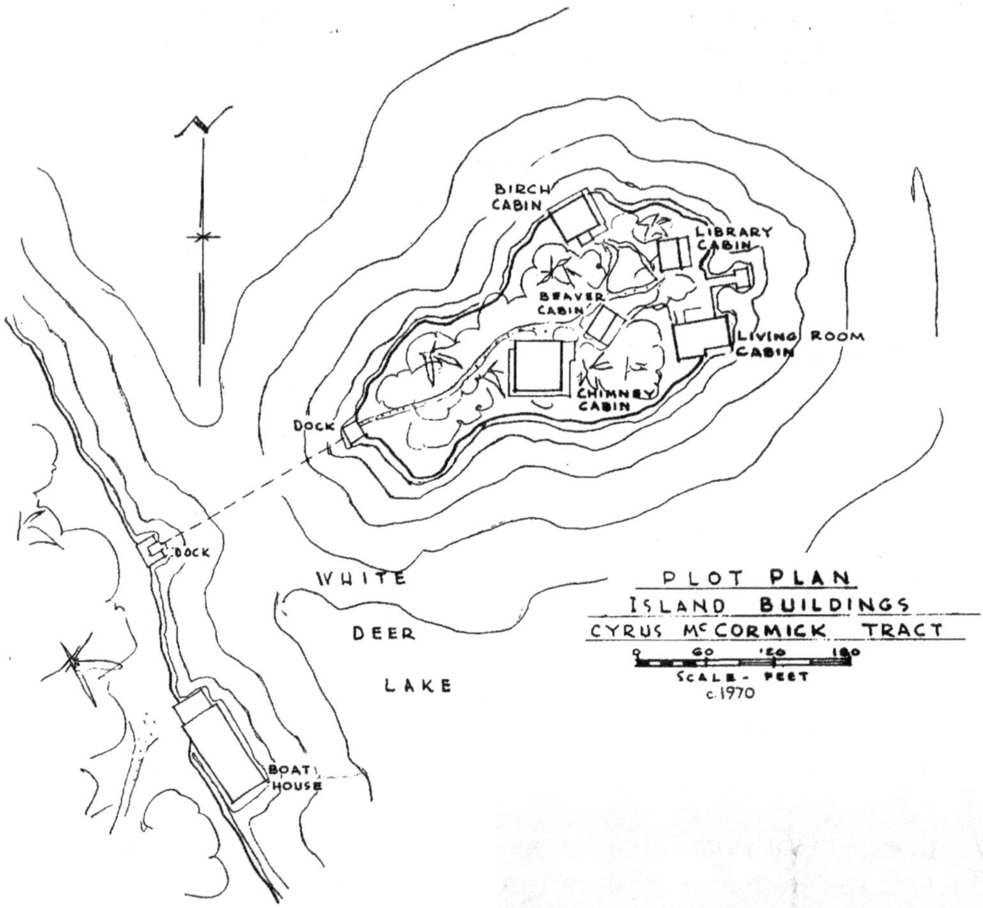

Above: Plot plan of the Island building, Cyrus McCormick Tract.

Right: A view of the Library Cabin, the first building built on the island.

*These photos and drawings are courtesy of US Department of Agriculture and Forestry Service.*

Cyrus McCormick was, in a way, forced to deal direct with Longyear, and in 1904, purchased the initial 151.75 acres for $3.16/acre, far less than the initial $5/acre first offered! The acreage purchased covered the west end of the lake as well as the island. Over the next 16 years, 13 different purchases expanded the property to 2,933 acres.

The first building on the island was built in 1904 and is named the library cabin. The other cabins on the island date about 1906, and the expansion on the lake shore was well developed by the 1930s. Bentley Cabin on Lake Superior was finished in 1905 and that same year the 26 mile-long trail, aptly named the Bentley Trail, (and still traversed today) was also completed.

In 1907, the lake was officially named White Deer Lake, in honor of the albino deer Cyrus and the others had frequently observed in the area.

October 1935 was the last time Cyrus visited the cabins on the island, as he passed suddenly on June 2, 1936, aged 77. The land was transferred to Gordon McCormick, son and architect, who began a renovation of all the buildings. The renovations lasted into the 1940s, Gordon's last visit to the settlement was 1947, possibly the same year the renovations were completed. When Gordon passed away in 1967, he willed the lands to the USDA Forest Service, with instructions that it was to become a wilderness tract, which it is today as part of the Ottawa National Forest.

The McCormick Wilderness Tract is now a National Forest and is preserved in much the same view as Cyrus would have enjoyed it in the 1890s. Camping and Hiking the area, as well as following the Bentley trail to Lake Superior, remains undisturbed today as it did over 100 years ago, with the exception of the stone foundations and a few posts and cables that remain, a subtle reminder that this wilderness remains remote as the time of the first explorers. Thanks to a bill passed in legislature designating wilderness areas, no impact camping is allowed and the use of motorized vehicles of any kind is not permitted. This area truly remains, and is, a remote wilderness.

The north portion of the tract contains virgin white pine and the Yellow Dog River Falls. With the headwaters of the Yellow Dog and Dead river beginning in the White Deer Lake, as well as many other streams, the vast remoteness of the native wilderness can be difficult to explore with numerous water falls, waterways and lakes. Old hiking trails made by the McCormick family can be found and used, revealing scenic overlooks and spectacular views of the Huron Mountains and nearby Lake Superior.

Above: A look at the screen porch which offered a comfortable place to enjoy the outdoors free from the black flies, mosquitoes and other insects that Michiganders are familiar with.
Below: Inside the Living room cabin and the relaxing atmosphere it invites.
*These photos and drawings are courtesy of US Department of Agriculture and Forestry Service.*

The buildings on the island and nearby shore of White Deer Lake were significant as it gave a wonderful example of a lodge resort that was popular at the turn of the century. The construction of the log buildings show a great attention to detail and the skilled craftsmanship of the period. The photographs and drawings are courtesy of the US Department of Agriculture, Forest Service, Marquette, Michigan, who obtained the following from the McCormick Estate around 1968. All that is left of the once magnificent buildings are a few stones from the foundation of the cabins on the island and nearby shore. The cabins were removed in 1970s and rebuilt elsewhere by a private owner.

Above: The view from the Birch Cabin Deck.
*These photos and drawings are courtesy of US Department of Agriculture and Forestry Service.*

Library Cabin Shows the appeal to the remote island, and the relaxing nature associated with a retreat at the turn of the century.
*These photos and drawings are courtesy of US Department of Agriculture and Forestry Service.*

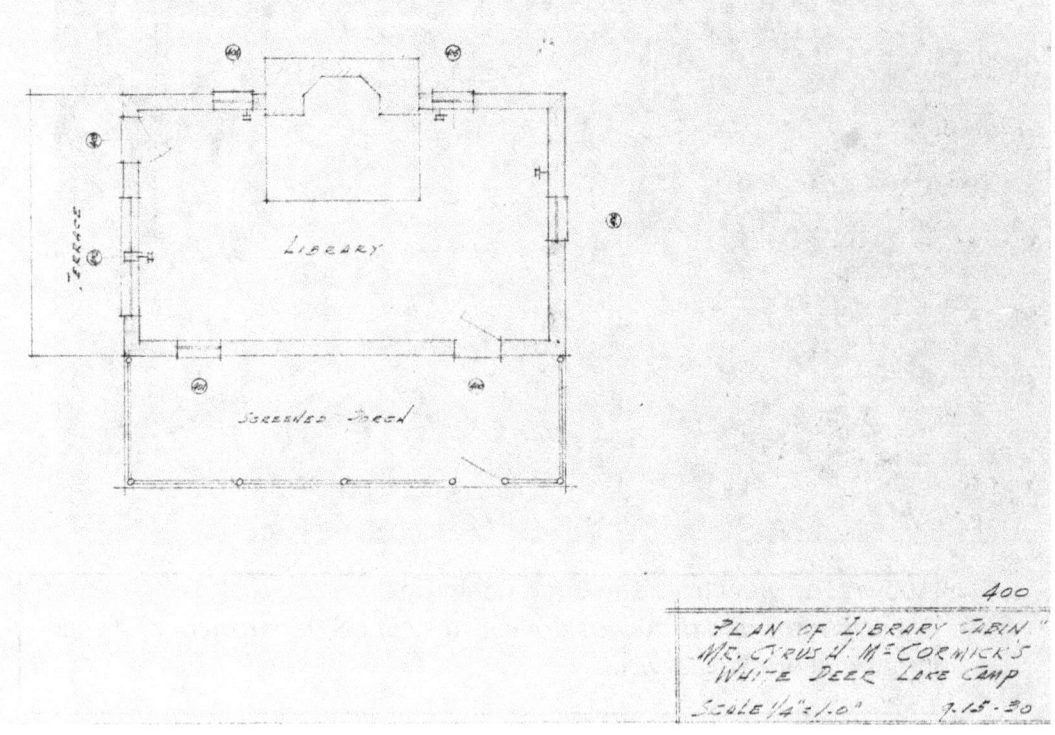

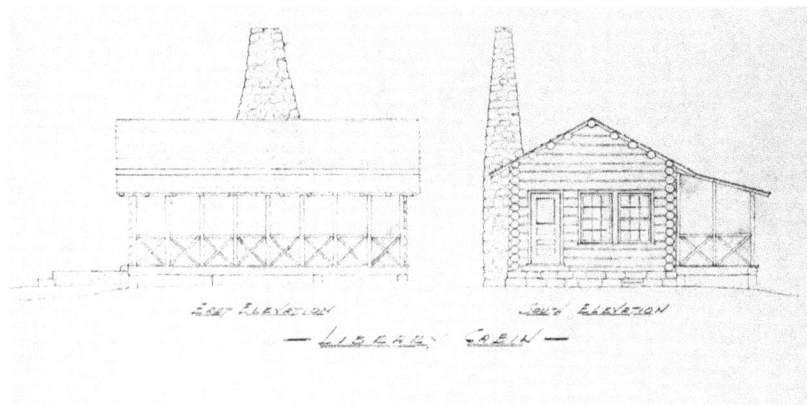

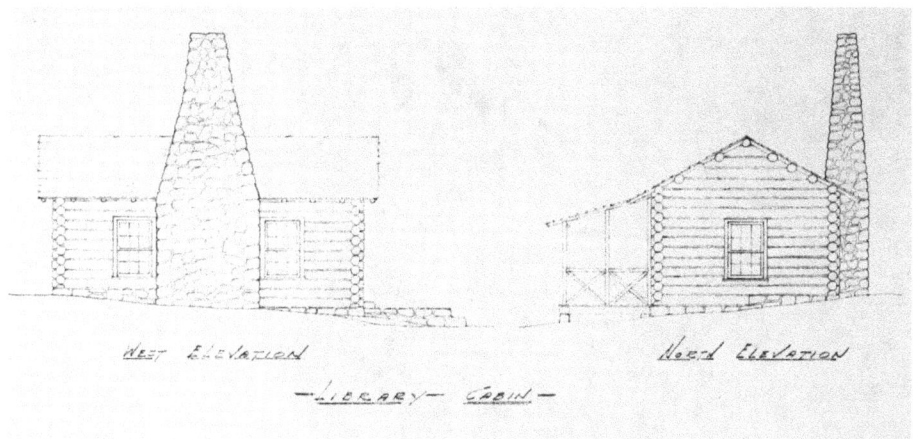

Above:
Elevation views of the Library Cabin

Right:
Relaxing sounds in nature accompanied by the soothing sounds of music.
*These photos and drawings are courtesy of US Department of Agriculture and Forestry Service.*

EAST ELEVATION
SOUTH ELEVATION
BEAVER CABIN
WHITE DEER LAKE CAM

WEST ELEVATION
NORTH ELEVATION
BEAVER CABIN
WHITE DEER LAKE CAM

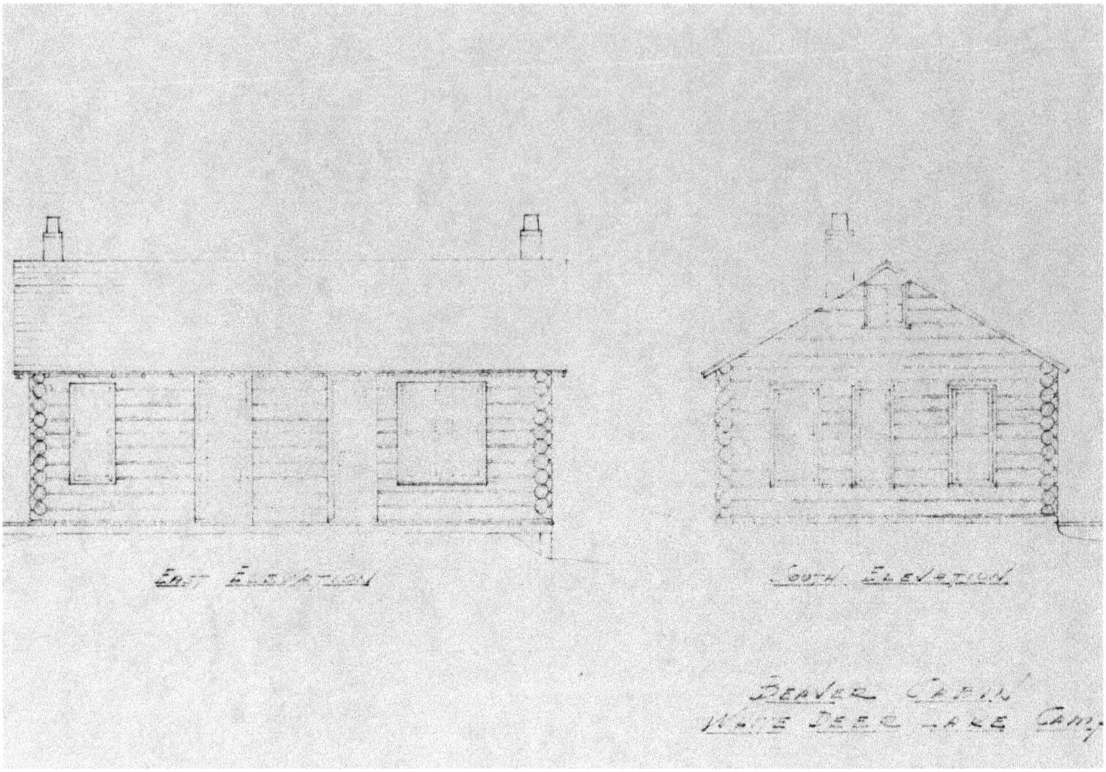

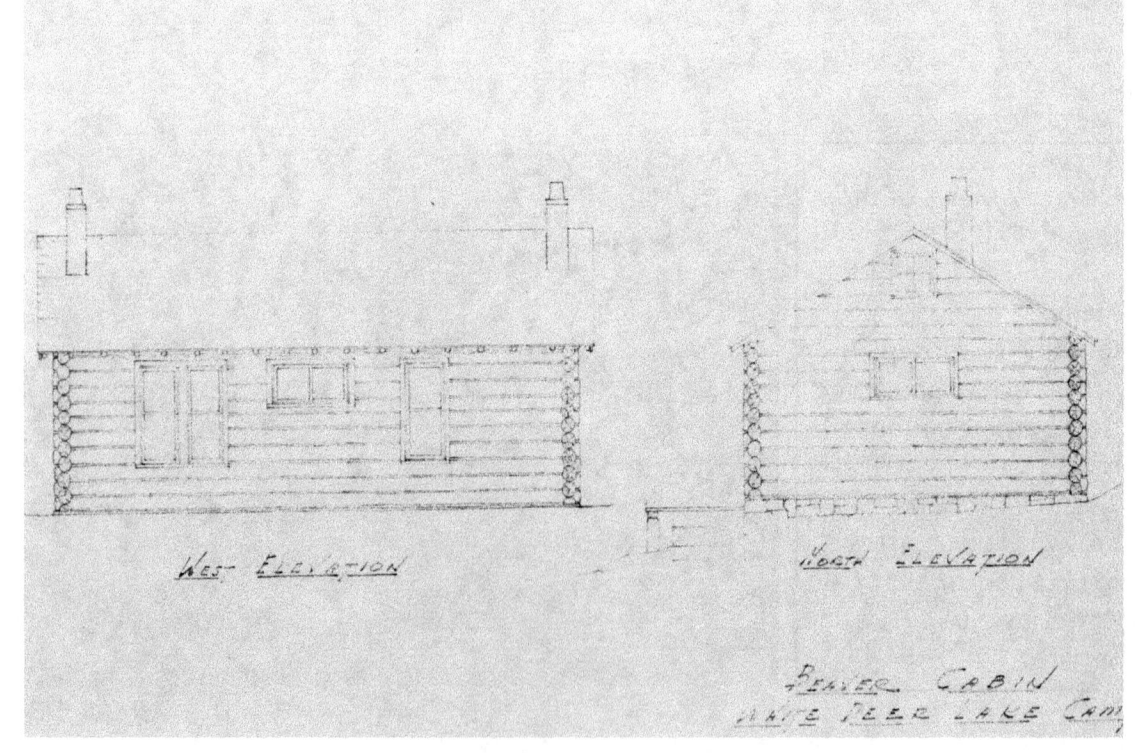

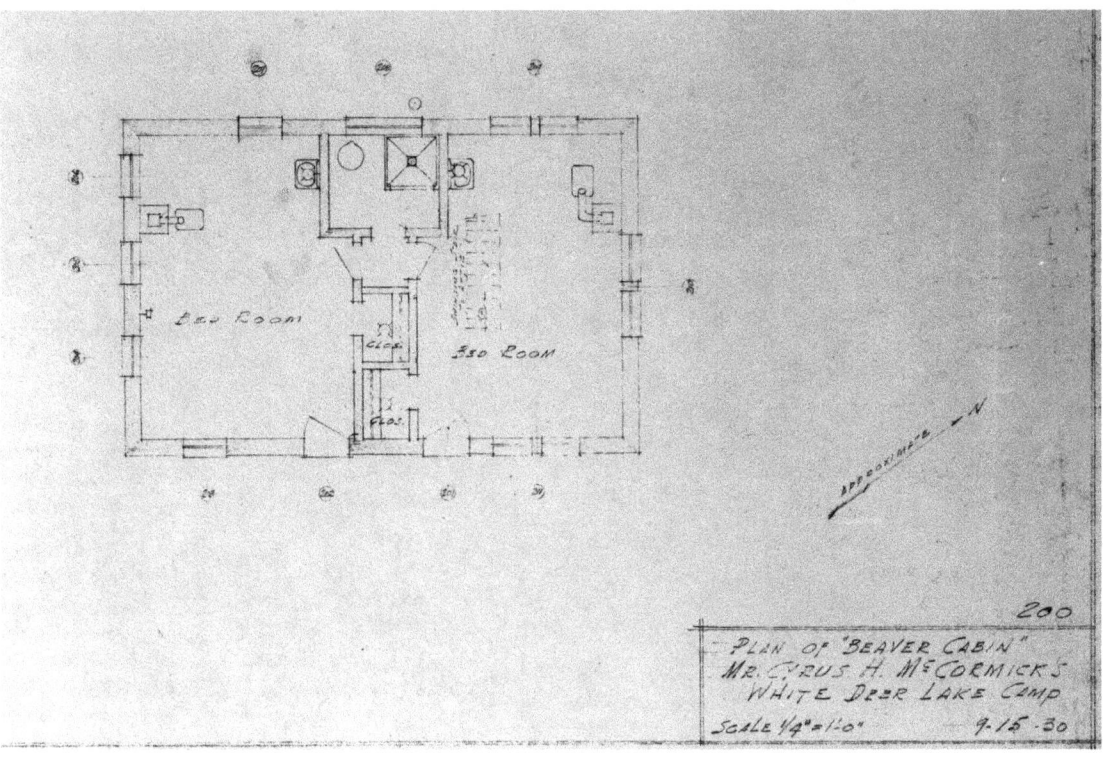

Beaver Cabin: A smaller cabin located on the island, near the Chimney Cabin. Left, Elevation views of the cabin. Above, Floor plan of the cabin. *The photos and drawings are courtesy of US Department of Agriculture and Forestry Service.*

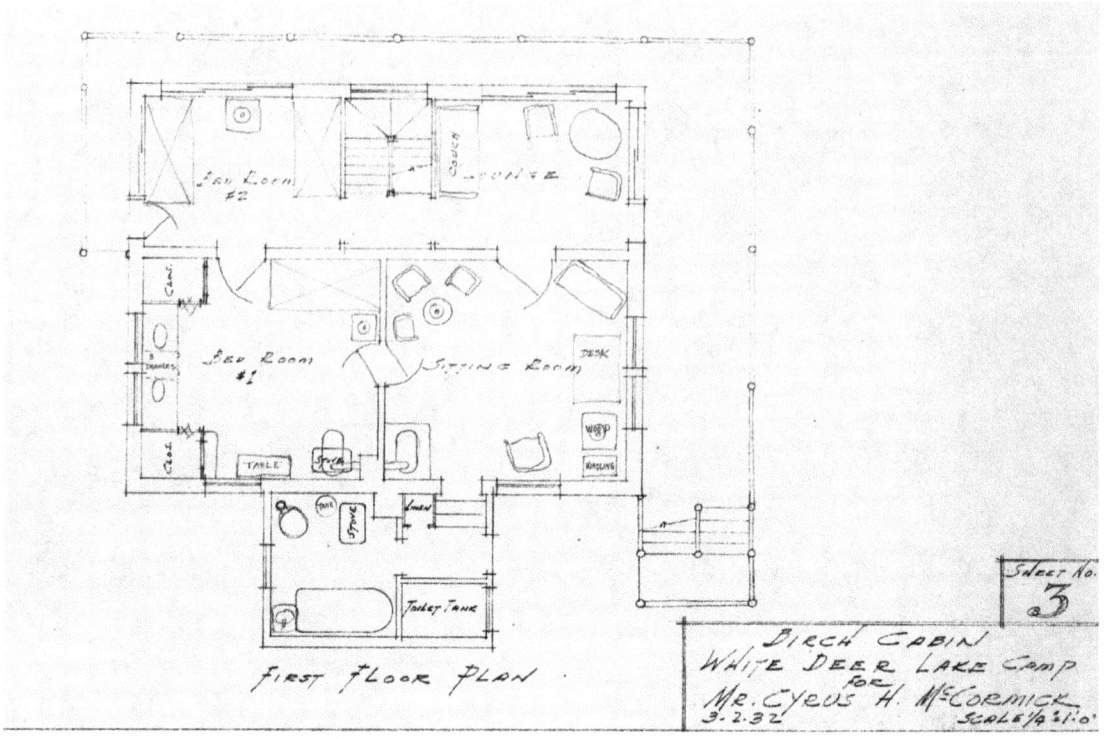

Above: First Floor plan of Birch Cabin.
Below: A view of the cabin from the side.
*These photos and drawings are courtesy of US Department of Agriculture and Forestry Service.*

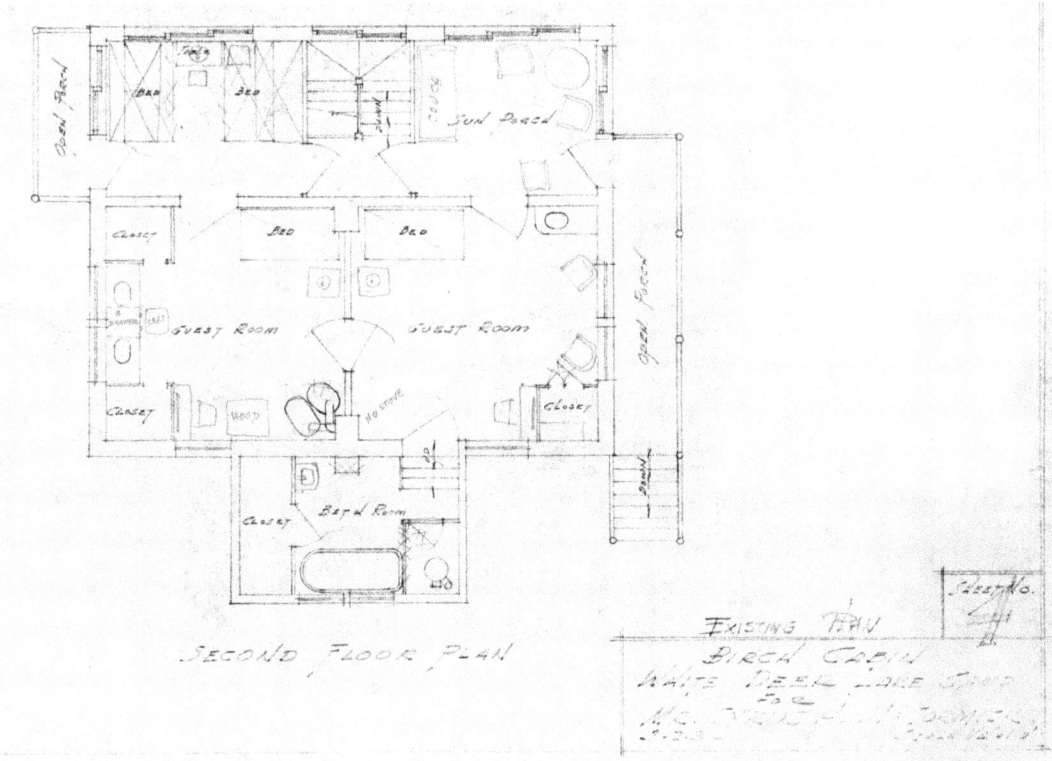

Birch Cabin, was also referred to as the Ladies' Cabin. It was a relaxing luxury hideaway with every need accounted for in the remote location. Above, Second Floor plan. Below, View of the cabin from the lake.
*The photos and drawings are courtesy of US Department of Agriculture and Forestry Service.*

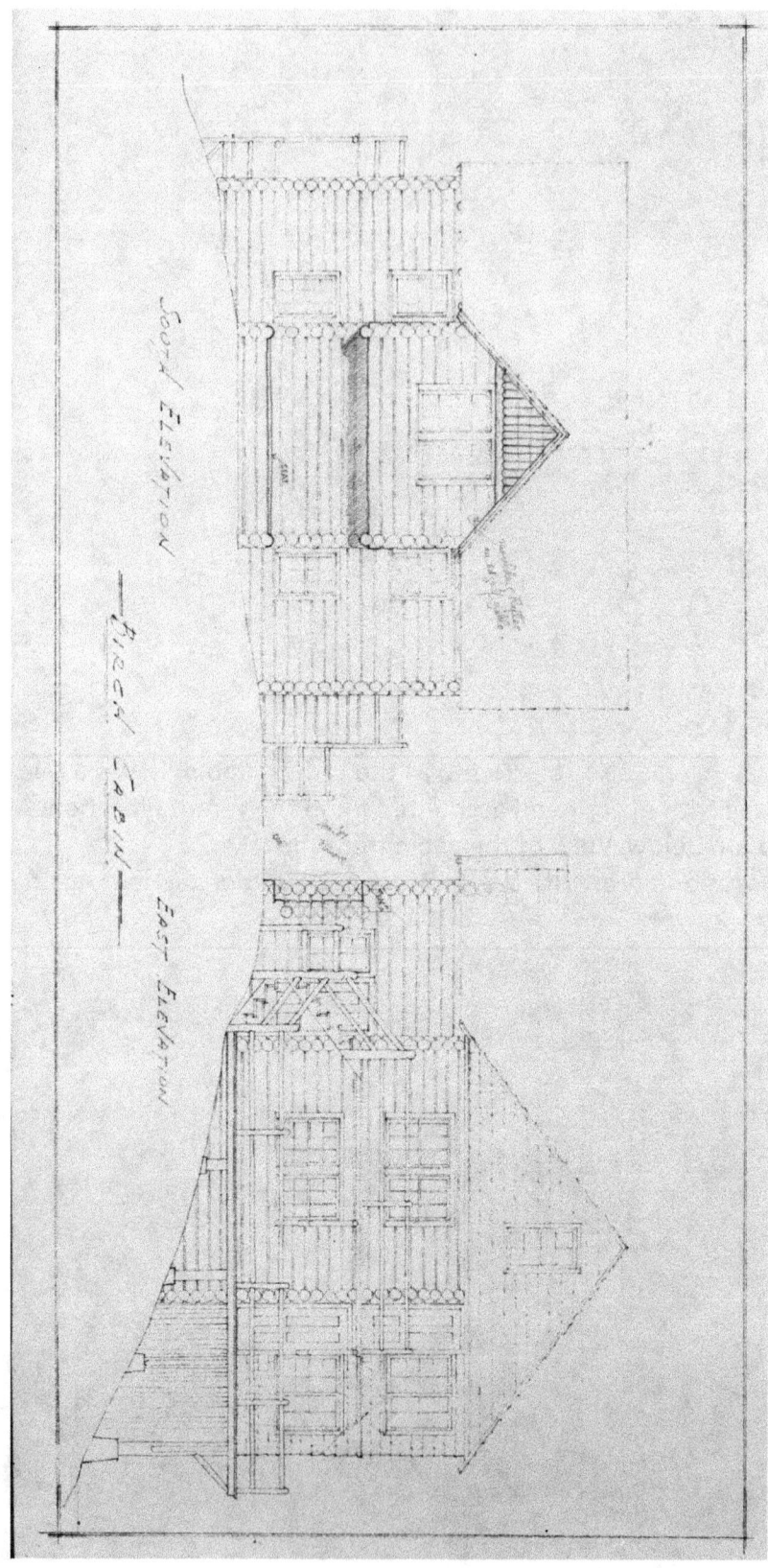

Left; Elevation plans for Birch Cabin,
Above; the cabins fit snugly into the surrounding forest on the island, including building the walkways and landings around existing trees.
*The photos and drawings are courtesy of US Department of Agriculture and Forestry Service.*

Living Room Cabin, Above, a favorite for gatherings with a cozy entertaining area and billiard room also had provision for boats to arrive and be protected in inclement weather.

Right: Floor plan for the Living Room Cabin.

*The photos and drawings are courtesy of US Department of Agriculture and Forestry Service.*

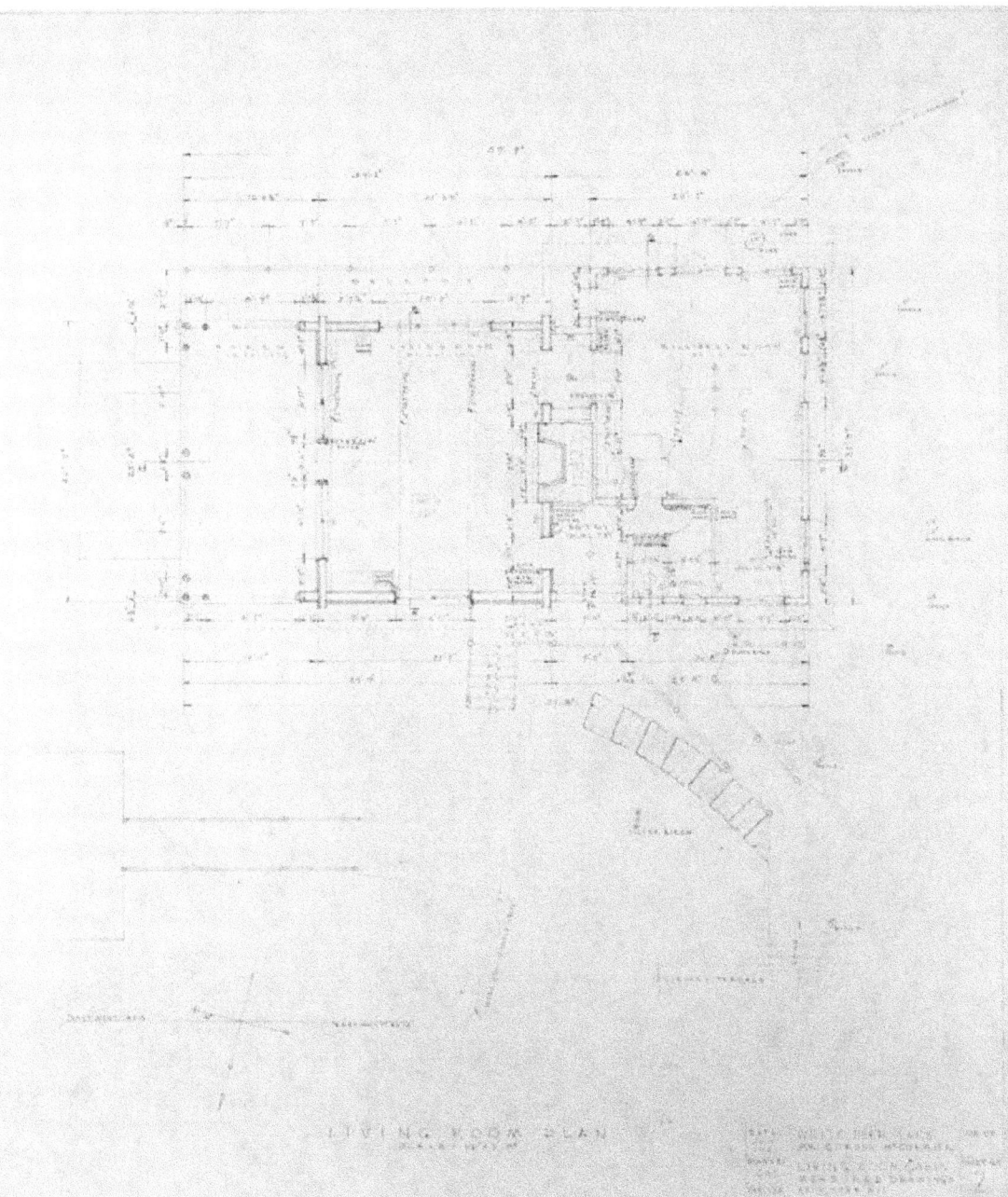

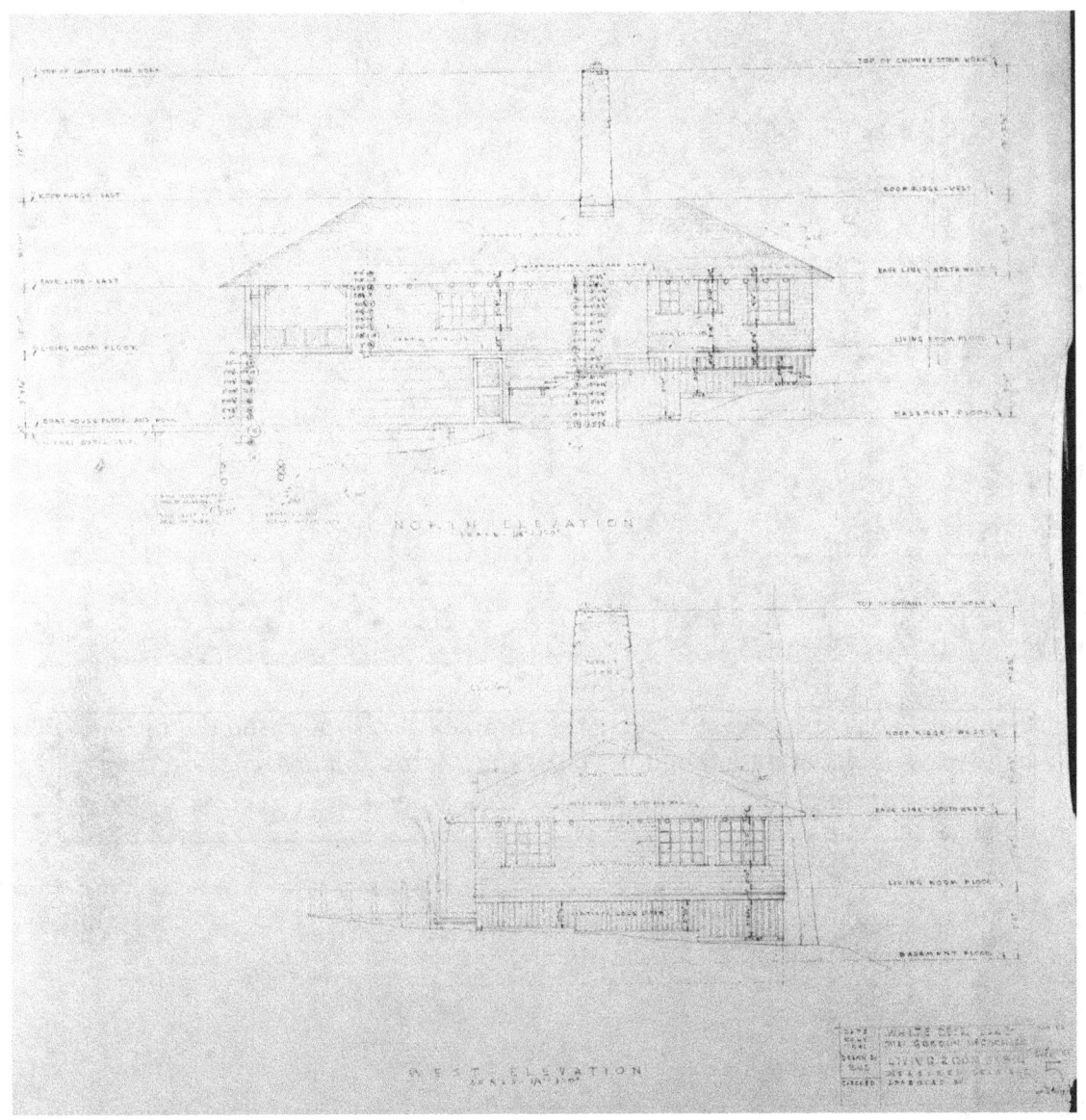

Living Room Cabin, a favorite for gatherings with a cozy entertaining area and billiard room also had provision for boats to arrive and be protected in inclement weather.
Left: South and East elevation plans
Above: North and West Elevation plans.

*The photos and drawings are courtesy of US Department of Agriculture and Forestry Service.*

Chimney Cabin, the largest cabin in the complex also showcased the most impressive views of nature and the lake. *The photos and drawings are courtesy of US Department of Agriculture and Forestry Service.*

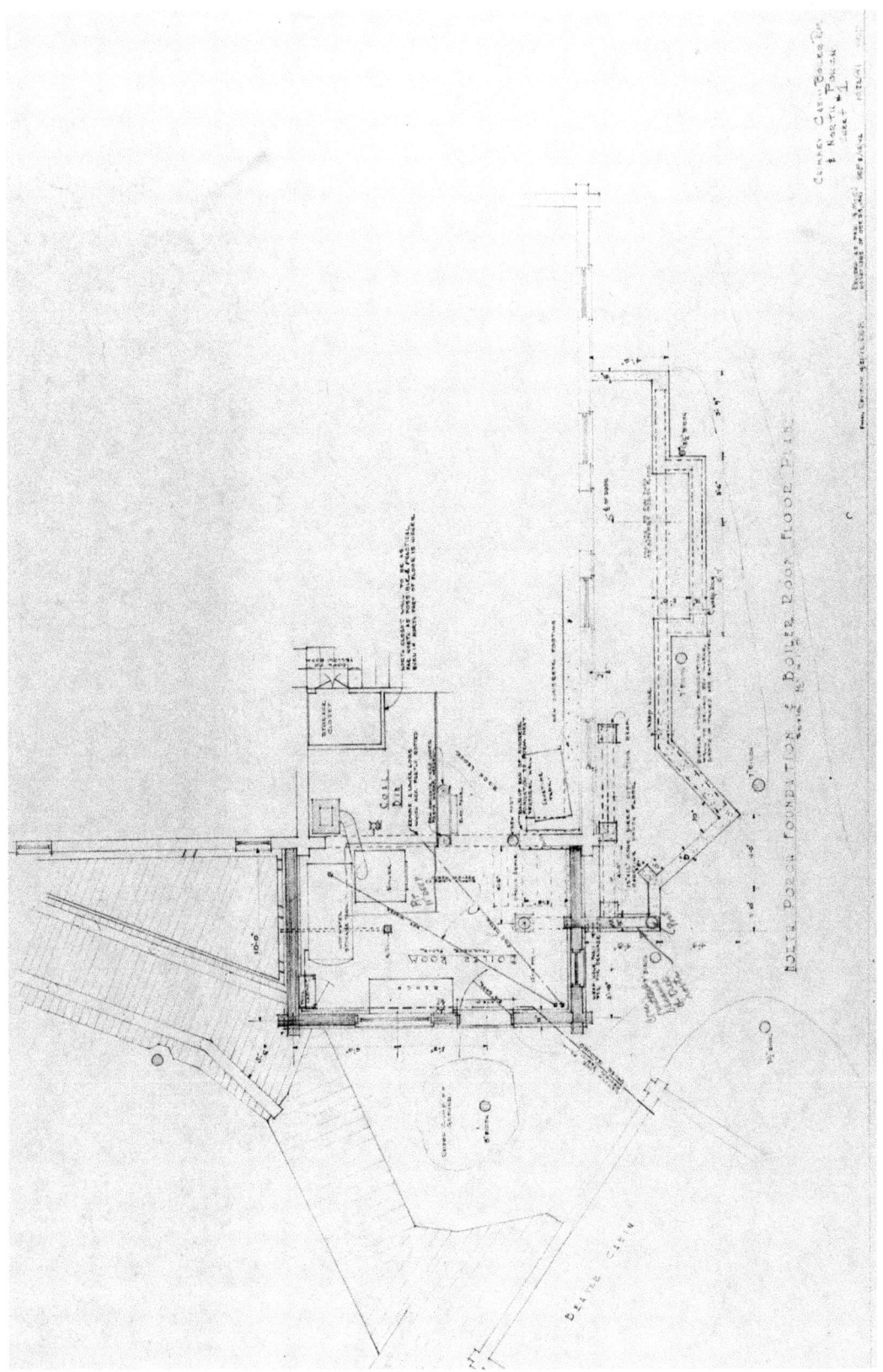

P

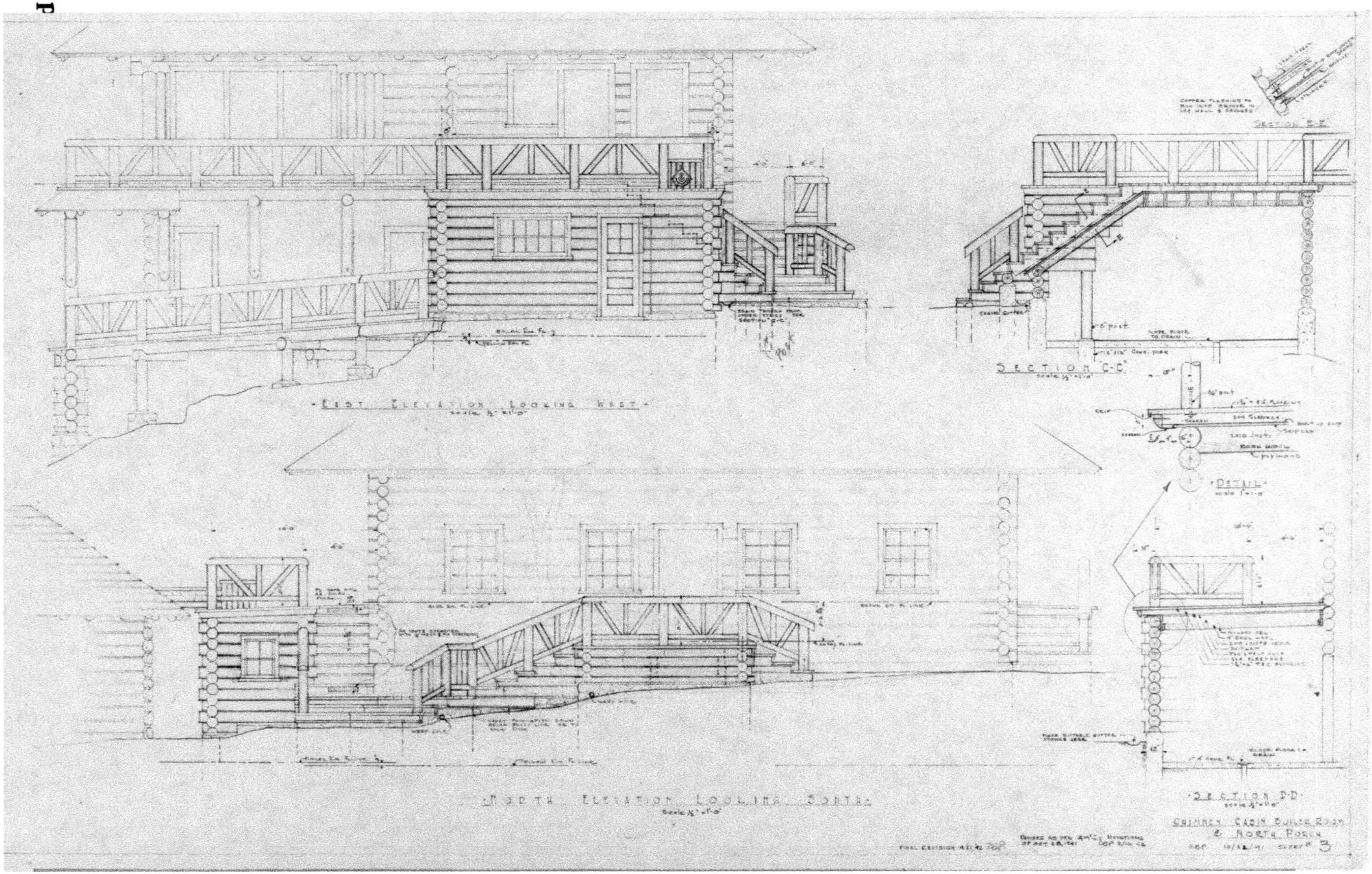
EAST ELEVATION LOOKING WEST
SECTION E-E
SECTION C-C
DETAIL
NORTH ELEVATION LOOKING SOUTH
SECTION D-D
CHIMNEY CABIN BUNK ROOM
& NORTH PORCH
SHEET # 3

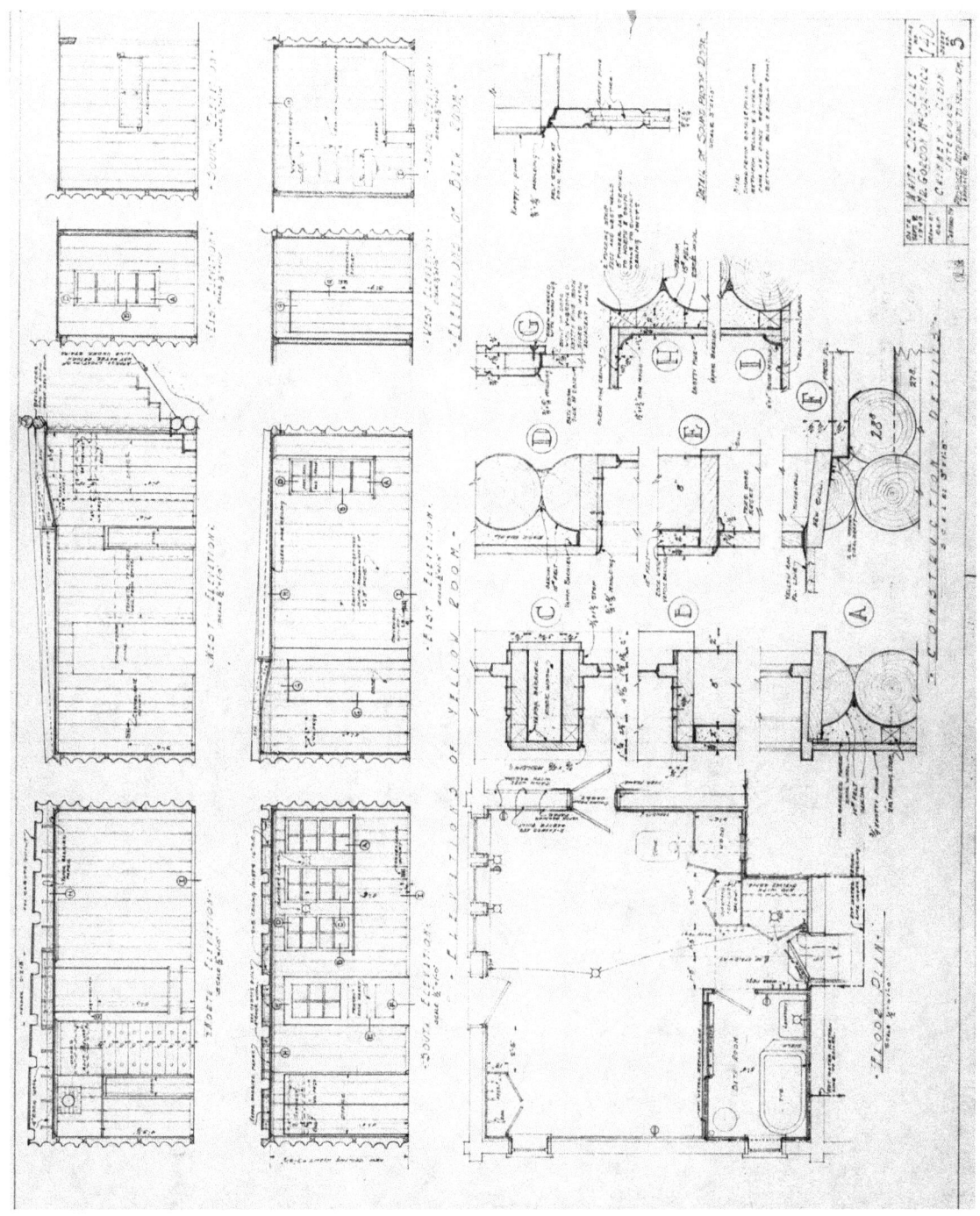

*Left: Chimney Cabin elevation views*
*Above: Inside room layout of Chimney Cabin.*
*The photos and drawings are courtesy of US Department of Agriculture and Forestry Service.*

SOUTH ELEVATION
CHIMNEY CABIN
SCALE 1/4"=1'0"
5-25-37
3


THIS EVE LINE LEVEL
THIS PORCH LINE LEVEL
EAST ELEVATION
CHIMNEY CABIN
SCALE 1/4"=1'0"
5-25-37
4

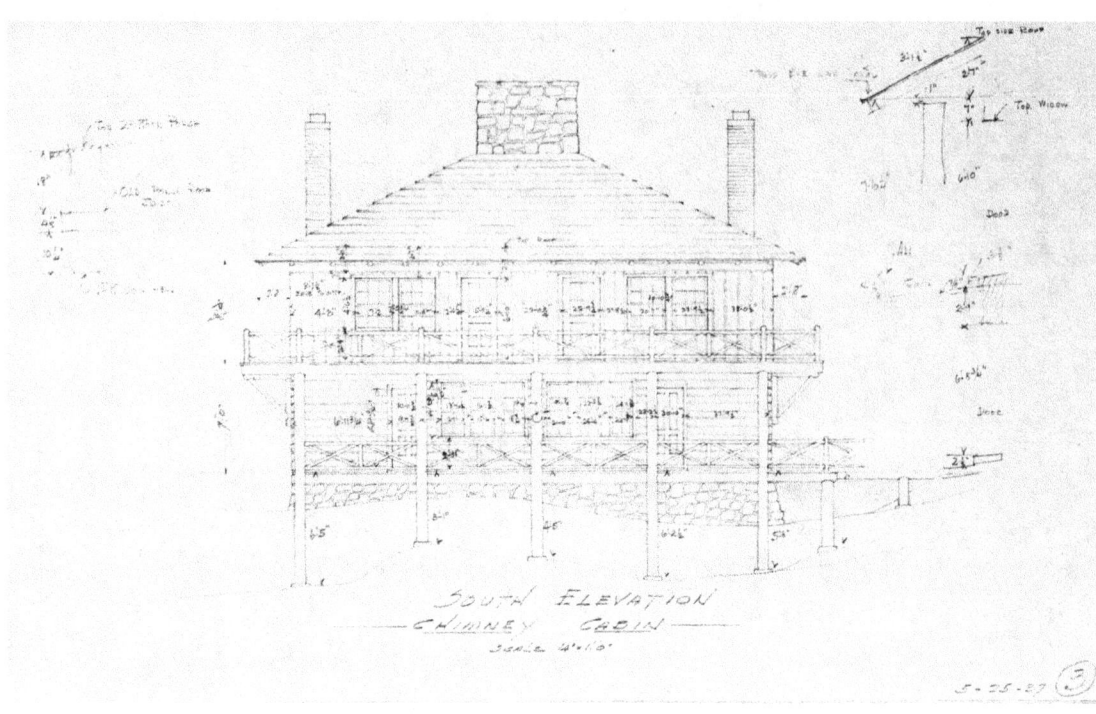

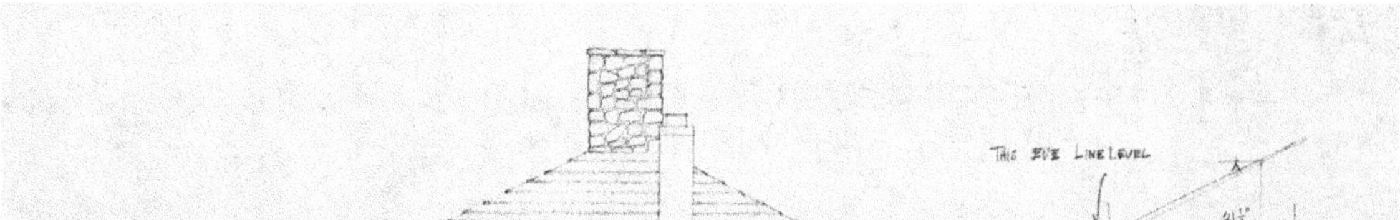

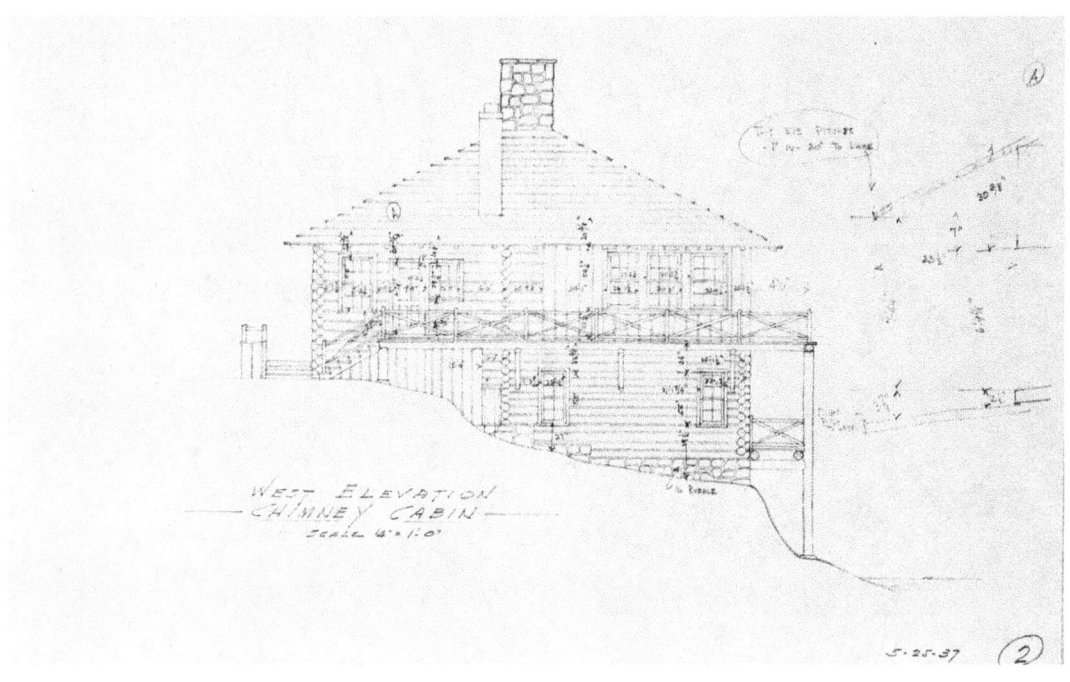

Left: South, East Elevation plans for Chimney Cabin.
Above: West, North Elevation plans for Chimney Cabin.
The photos and drawings are courtesy of US Department of Agriculture and Forestry Service.

Elevated view of the headquarters of the International Harvester Company at 180 North Michigan Avenue. The building was completed in 1937 and was located at the intersection of Michigan and Lake. WHS #6631

## GENERAL VIEWS OF HARVESTER DEALERSHIP BUILDINGS

On the following pages are different views of International Harvester Dealerships, a short history is given if available, with location.
One thing that stands out is the unity that the dealership presented, throughout every state, in every town, International Harvester was a recognized name. From Home canning and preservation demonstrations, to tractor schools and tractor repair classes.

Farmers Union Dealership in Columbus, Kansas
Demonstration of correct Meat wrapping and storage for refrigerators and Freezers. 1948 era.

*Photos courtesy of B. Bauer, Kansas.*

A view of Columbus, Kansas
Taken in 1956 off the top of the grain elevator which is next to the Farmers Union Co-op dealership, looking west down Maple street, Columbus, Kansas. The building on the left with the big sliding door open is Long Bell lumber company; originally started in the 1870s, and eventually turned into the International Paper Corporation.
*Photos courtesy of B. Bauer, Kansas.*

Farmer's Union Coop. Columbus, Kansas; looking west and east on East Walnut street, 1950s.
*Photos courtesy of B. Bauer, Kansas.*

O'Keefe Implement was established in the year 1916 at Yale, South Dakota by J.E. O'Keefe and his brother Pat. Due to the depression of the "Dirty Thirties" J.E. O'Keefe moved his farm equipment business 26 miles east to the Pearson's old livery barn in DeSmet. This livery barn is mentioned in the "Little House on the Prairie" Laura Ingalls Wilder books.

In 1990, the old Pearson livery building was torn down and a new building was constructed. Today the business still operates from the same main street DeSmet location since the 1930'S.
O'Keefe Implement celebrated its 100th Anniversary in 2016

*Courtesy of Howard Raymond.*

In 1915, Lewis Scott purchased the Helgerson Implement building in Mt Vernon, South Dakota, which was already an International Harvester dealership. In these first years, Lewis continued selling automobiles as well as gasoline, farm supplies, a full line of farm machinery, and Titan tractors.

Wally Scott joined his father in 1936. Scott Supply Co. moved to Mitchell in 1962. Current owners Robert Scott started in 1963, John Scott started in 1973, and Chris Scott started in 1994.

Scott Supply celebrated its 100th Anniversary in 2015

*Courtesy of Howard Raymond.*

Martin Hersrud opened a general store in Petrel, North Dakota, two miles from his homestead in 1908.

In 1910, he started selling farm implements, which were all horse-drawn.

In 1911, Martin started offering McCormick-Deering implements for the International Harvester Company.

In 1918, he moved the store about 10 miles from Petrel to White Butte, South Dakota, and was there for probably five years before a fire completely destroyed the building his store was in. At that point he moved the business to Lemmon, South Dakota.

In the early 1940s, sons Morris and Gordon, joined the business, and eventually took over the dealership upon Martin's retirement in 1951.

Today, David Hersrud and his brother-in-law Arnie Luptak represent the third generation of the Hersrud family operating the business out of Sturgis, which has grown to include a second location in Belle Fourche, over 50 employees and a 100-year history of meeting the area's automobile and farm equipment needs.

Hersruds is currently the oldest continually running family owned dealership in western South Dakota, and celebrated 100 years of business in 2010.

*Courtesy of Howard Raymond.*

Inside view of the 1913 International Harvester Almanac *(Tomac)*

## If People Haven't the Means, They Cannot Purchase

There is no more effective way of increasing sales than to increase the buying power of the farmer and this can only be done by increasing farm production.

The International Harvester Company maintains a special department to promote better farming methods and greater farm production.

The Agricultural Extension Department was organized six years ago. Its director is Professor P. G. Holden, known the county over, by reason of his efficient work with the University of Illinois and the Iowa State Agricultural College.

The only purpose of the Department is to help the people on the farms and in the homes to improve their conditions, industrially, financially, and socially.

Every state, every community has its own problems directly affecting the farm, the home or the school, and the Extension Department is ready at any time to send trained experts into any community to help the people solve the problems that confront them.

Whenever any community needs and desires their assistance, these lecturers, educators, demonstrators, and practical farmers respond promptly. They hold meetings in halls, in churches, in schoolhouses, in homes, in empty store buildings, in tents, in vacant lots, or out on the farms - wherever people can gather to hear them. And they discuss the problem that is most important to that community. it may be diversified farming; it may be livestock or dairying. It may be to show people how to feed themselves, or how to build up the fertility of their soil. It may be to discuss corn, oats or wheat or alfalfa or poultry. Possibly it is animal diseases, plant diseases, or insect pests. It may even be how to get rid of the death-dealing fly, how to grow gardens, how to can or dry or store fruits and vegetables. it may be sanitation or home economics or good roads or home ownership or community development.

During the six years the Department has conducted campaigns and short-course schools in nearly every state in the Union. Its members have traveled approximately a million miles by train and over a quarter of a million miles by automobile, held over fifteen thousand meetings and addressed 2,000,000 people.

What have been the results? Here is just one of them: on June 11th of the present year (1919), at a dinner given at the Union Stock Yards to a visiting delegation of Arkansas bankers, business men, and farmers, Governor Charles H Brough of that State said:

"We feel a big debt of gratitude to the International Harvester Company - a personal obligation, when we are reminded that in 1914, when the average productive power of the Arkansas farmer was $169 per year, that Company sent a man into the state of Arkansas, who enabled us to increase our agricultural production per farmer until it is now $1,320 per year. I refer to our friend, your great agricultural representative, Professor P. G. Holden. This great work of profitable farming is expected to increase production of our State by $300,000,000 in the next five years, and give the farmers a margin of $60,000,000 a year."

Thus a governor publicly voiced the lasting good will of the 1,700,00 people of his great State for the Harvester Company and credited this Company with having increased the purchasing power of the State over ten times.

To increase production, eliminate waste, improve living conditions in town and country, give more prosperity to the farmer, make housework easier for the farmer's wife, provide better schools for the farm boy and girl and create within them a love for the farm and greater interest in agriculture - in brief, to make better farms, better homes, better schools happier families - these are the things to which the Agricultural Extension Department gives its entire attention.

For these things mean more efficient people, greater production, greater prosperity.

No other implement concern has applied such broad principles to its business - increasing the future buying power of the ultimate purchaser by increasing his prosperity and efficiency. And the application of this principle works just as much to the dealer's advantage as it does to our advantage. Every International dealer can avail himself of the Agricultural Extension Department's farm educational service, which not only eventually increases the farmers buying power, but also gives the dealer added prestige in his community by reason of having been identified with such a worthwhile movement.

M.R. Mosher.

*Transcribed Direct from the Memoir of P.G.Holden, 1944.*

*This letter is a great example of the enthusiasm that PG Holden carried with him throughout his career and life. To encourage the farmer, the student, or the community, he challenged them to "Beat their Best" and be a better person than they were the day before. Neat buildings and inviting appearance are the first step to a prosperous and profitable farm.*
*Without Holden, Cyrus McCormick Jr. would not have been able to carry on his mothers vision of a family company, where the 'family' was any one who was interested in the advancement of agriculture. Holden grew up having to complete household chores, in a time where that was uncommon. He clearly had a great understanding of every aspect that it took to operate a successful family farm and he carried that with him wherever he went.*

Modern Farm Buildings on "Sedgeley" Dairy Farm, Near Hinsdale, Ill.

Acknowledgments and Resources

International Harvester Almanacs, 1909-1919

Osborne Annual Almanacs, 1912-1914

International Harvester Service Bureau,
    Multiple publications, noted in text, 1912-1944

McCormick, Cyrus. The Century of a Reaper, 1931

Holden, Perry Greely. Memoir Holden and Wilson Families, 1944

Roderick, Stella Virginia . Nettie Fowler McCormick
            Rindge, N. H. : R. R. Smith, 1956

Rosenburg, Chaim M..The International Harvester Company:
            A History of the founding families and their machines, 2019

Nelson, Daniel. Farm and Factory: workers in the midwest 1880-1990, 1995

Thanks also is due to the following:

Kenneth Updike

Wisconsin Historical Society

T. Clark at Navistar

Also, untold levels of gratitude is owed to fellow International Harvester collectors, former International Harvester workers, and current members of the International Harvester Spirit. (That's you, my readers) Without you, the Spirit of the Harvester family would be lost. Thank you for keeping the innovation, prosperity, efficiency and dedication of the agricultural revolution at the forefront, raising the most important crop of all, our children. Thank you.

www.ingramcontent.com/pod-product-compliance
Lightning Source LLC
Chambersburg PA
CBHW081748100526
44592CB00015B/2338